Unlocking the Power of Auto-GPT and Its Plugins

Implement, customize, and optimize Auto-GPT
for building robust AI applications

Wladislav Cugunov

Unlocking the Power of Auto-GPT and Its Plugins

Associate Group Product Manager: Niranjan Naikwadi

Publishing Product Manager: Nitin Nainani

Book Project Manager: Aparna Nair

Senior Editor: Tiksha Lad

Technical Editor: Rahul Limbachiya

Copy Editor: Safis Editing

Proofreader: Tiksha Lad

Indexer: Hemangini Bari

Production Designer: Vijay Kamble

Senior DevRel Marketing Executive: Vinishka Kalra

First published: September 2024

Production reference: 1160824

Published by Packt Publishing Ltd.

Grosvenor House

11 St Paul's Square

Birmingham

B3 1RB, UK.

ISBN 978-1-80512-828-1

www.packtpub.com

To my parents and family, who have observed me sticking my nose into my computer display since I was 9 years old; my friends, who tolerated my passion; and my wife, who supported and helped me while I worked on several projects, including this book.

– Wladislav Cugunov

Contributors

About the author

Wladislav Cugunov is a senior software developer and visionary technologist known for his exceptional skills and innovative, determined mindset. At age 12, he built a LEGO helicopter to prove to his dad that LEGO could indeed fly, even though it exploded after he doubled the electricity. At 15, he constructed his first Tesla coil, causing an **electromagnetic pulse** (**EMP**) when attempting to drive a coilgun with it.

His coding journey began with Batch and VBA when he tried to replicate Jarvis from *Iron Man* in 2008, progressing to Node.js to build a trading bot, Bash, Java, Ruby, C, Python, Kotlin, C#, PHP, and Swift. Dissatisfied with the direction of Auto-GPT, he started making suggestions and improvements until he was invited to come on board and help AutoGPT grow.

I extend my deepest gratitude to my family and wife for their unwavering support and encouragement. Special thanks to my friends David and Alex, who inspired me to move forward so that I could inspire them as well. I also appreciate the developers and contributors to the open source community, whose innovations and dedication have guided my work. Lastly, thank you to my readers for their interest and trust in this journey.

About the reviewers

Venkatachalam Rangasamy, an innovator with over 18 years of experience, has transformed the landscape of software development. With a start-up mindset, he has bootstrapped numerous groundbreaking projects, earning industry accolades for his innovative solutions. Leading global engineering teams, Venkatachalam has driven the development of AI-driven platforms that significantly enhance operational efficiency and customer satisfaction. Passionate about generative AI, he is developing a platform to seamlessly connect enterprise data and make software development cheaper and faster.

His creative spirit shines through his 25+ patents and industry-wide presentations, continually pushing technological boundaries.

Sajid Patel, recipient of a bachelor of engineering in mechanical engineering from Gogte Institute of Technology, is a global digital manufacturing development manager with over 26 years of experience in data science, software/application development, and performance engineering. He is a self-directed, hands-on leader with expertise in **Internet of Things** (**IoT**), machine learning, **product lifecycle management** (**PLM**), optimization, and Solidity/Blockchain. He has a proven track record of success in various industries, including product development/R&D and software services/consultancy. His experience spans sectors including automotive, aerospace, and manufacturing. Sajid is proficient in multiple programming languages, including JavaScript, Python, R, Java, C, and C++.

Purusoth Mahendran is a seasoned engineering manager with over four years of management experience and over nine years in developing and scaling large-scale,data-sensitive systems. Currently, he serves as a senior engineering manager at Thumbtack, driving strategic initiatives and leading cross-functional teams. At Cash App and Amazon, Purusoth built advanced machine learning solutions that enabled innovative product features. Known for his visionary leadership, he excels in fostering innovation and aligning technical strategies with business goals. He holds a master's in computer science from the University of Texas at Dallas and a bachelor's of engineering degree from the College of Engineering Guindy in India.

Tanmaya Gaur is a principal architect at T-Mobile US, Inc., with over 15 years of experience in building enterprise systems. He is a technical expert in the architecture, development, and deployment of advanced software and infrastructure for enhanced user support in telecom applications. He is passionate about utilizing composable architecture strategies to aid the creation and management of reusable components across the entire spectrum of web development, from frontend UX to backend coordination and content management. His current focus is building telecom CRM tools that leverage micro-frontend, artificial intelligence, and machine learning algorithms to boost efficiency and improve the product experience.

Vishal Soni is a senior-year student at the **Indian Institute of Information Technology (IIIT)** Bhagalpur, specializing in computational methods for solving real-world challenges. As a student researcher, he has worked at top labs such as the Swiss AI Lab IDSIA in Switzerland and NUS Singapore, contributing to cutting-edge AI models. He has also conducted research at IIT Roorkee and IIM Bangalore, focusing on AI, data analytics, and optimization. He is proficient in MATLAB, machine learning, AI, and soft computing. Vishal regularly participates in courses and seminars to stay at the forefront of his field. A former member of IEEE and SCRS, he has had two first-author publications accepted at prestigious conferences in India and Spain.

Table of Contents

6

Scaling Auto-GPT for Enterprise-Level Projects with Docker and Advanced Setup   71

7

Using Your Own LLM and Prompts as Guidelines   85

Preface

Thank you for picking up this book. I hope it will serve you and help you have fun with Auto-GPT or even improve your or other people's lives!

Every innovation story is intertwined with moments of serendipity, and mine began while I was immersed in building my own orchestrator for ChatGPT. The intention was to create a tool that could facilitate seamless conversation between multiple instances of ChatGPT, one of which would act as an "instructor" capable of spawning new instances and delegating tasks to them. To achieve this, I was working on a JavaScript injection mechanism.

Despite the technical challenges—primarily the complexity of conveying the right context to newly cast instances—I remained undeterred. The journey was not easy, especially since this was before the arrival of GPT-4, a more advanced model capable of grasping context more efficiently.

It was during this time of intense research and testing that I came across an episode of *Tech Quickies* by *Linus Tech Tips*, discussing HustleGPT. Intrigued by this newfound information, I plunged deeper into researching it. This led me to stumble upon Auto-GPT—an AI project remarkably similar to the orchestrator I had envisioned.

Auto-GPT, with its ability to interact via the OpenAI GPT API to recursively solve tasks and operate in a browser environment, was a marvel to behold. More interestingly, it incorporated a plugin system, extending its capabilities even further. As a freelance software developer with a deep-seated passion for AI, this discovery was a watershed moment in my career.

What began as an academic interest in Auto-GPT soon morphed into a desire to be a part of its development. I found myself spending a considerable amount of time understanding, tinkering with, and, eventually, contributing to the project.

My most substantial contributions lie in the realm of Auto-GPT plugins and **text-to-speech** (**TTS**) functionality. I have always believed in the power of AI to not only interpret text but also deliver it in the most natural and human-like manner. Enhancing the TTS capabilities of Auto-GPT, I aimed to bridge the gap between human and machine interaction.

Alongside this, my relentless pursuit of innovation has led me to ideate, develop, and integrate new features and improvements into the Auto-GPT code base. Being part of the project, I also recognized the need for swift and efficient code review processes. I committed myself to act as a catalyst, meticulously reviewing pull requests, offering constructive feedback, and consistently refining my own code.

The culmination of these efforts is this book, *Unlocking the Power of Auto-GPT and Its Plugins*. More than a chronicle of my experiences, it serves as a comprehensive guide to understanding and harnessing the power of Auto-GPT. Through this book, my goal is to do the following:

- Introduce the fundamentals of Auto-GPT
- Guide you through the installation and setup of Auto-GPT
- Teach the art of crafting effective prompts
- Demystify the usage and customization of Auto-GPT plugins
- Foster creativity in developing AI applications
- Detail advanced setup with Docker
- Enlighten on the safe and effective use of the continuous mode
- Help you integrate your own language model with Auto-GPT

Join me on this exciting journey as we delve deeper into the fascinating world of Auto-GPT. Together, let us unravel the immense potential of AI and redefine the future of technology.

Who this book is for

This book is designed for a wide range of readers, from AI enthusiasts and hobbyists to professionals in software development, AI, and data science. Whether you are a beginner with no prior experience in AI or an expert looking to explore new applications of AI technology, this book has something for you.

Specifically, this book is for the following:

- Software developers interested in integrating AI capabilities into their applications
- Data scientists eager to leverage AI for data processing and analytics
- AI hobbyists looking to experiment with creative applications of AI technology
- Students pursuing degrees in computer science, AI, or related fields
- Educators who wish to incorporate practical AI applications into their curriculum
- Business professionals interested in understanding how AI can optimize business processes
- Non-technical individuals curious about the potential of AI technology

What this book covers

Chapter 1, *Introducing Auto-GPT*, offers an overview of Auto-GPT, its history, and its place in the AI ecosystem. You will learn what Auto-GPT is and why it is a revolutionary technology.

Chapter 2, From Installation to Your First AI-Generated Text, guides you through the initial setup and installation of Auto-GPT. You'll learn how to get started with the basics and set yourself up for success.

Chapter 3, Mastering Prompt Generation and Understanding How Auto-GPT Generates Prompts, delves into the mechanics of how Auto-GPT generates responses to prompts. This will help you understand how to effectively communicate with the AI and get the results you're looking for.

Chapter 4, Short Introduction to Plugins, introduces you to the world of plugins in Auto-GPT. You will learn what plugins are, how they can extend the functionality of Auto-GPT, and how to get started with using them.

Chapter 5, Use Cases and Customization through Applying Auto-GPT to Your Projects, gives practical examples of how to use Auto-GPT for various tasks, and how to customize it to suit your needs. This chapter will include case studies and real-world applications.

Chapter 6, Scaling Auto-GPT for Enterprise-Level Projects with Docker and Advanced Setup, covers the continuous mode feature in Auto-GPT and discusses its implications. You'll learn how to use this feature effectively and safely, and you'll learn how to set up Auto-GPT in a Docker environment. This is for advanced users who want to create a more controlled and scalable setup for Auto-GPT.

Chapter 7, Using Your Own LLM and Prompts as Guidelines, guides you on how to integrate your own **large language model** (**LLM**) with Auto-GPT. You will understand the steps involved and the potential benefits of using a custom model, and you will learn the art of crafting effective prompts. You will learn how to guide Auto-GPT in generating the responses you want, and how to maximize the quality and relevance of its output.

To get the most out of this book

Here are a few things you need to understand this book better:

- **Have a basic understanding of AI**: While AI is pretty advanced now, you will still encounter many cases where the results you get just do not make sense. For that, you will need a lot of patience.

- **Using GPT costs money**: While there are free LLMs in the wild, Auto-GPT focuses on OpenAI's GPT-4 mostly and runs best with it. Running Auto-GPT costs may differ, so use it at your own responsibility and set usage limits to keep your expenses under control.

- **Install the necessary tools**: Make sure to install the tools that we will use in *Chapter 2*; skipping the installation will only set you back.

- **Research, research, research**: While troubleshooting can be done fairly quickly in most cases, it can be hard to understand issues when running Auto-GPT, especially if you do not have any programming background. The best you could do is use Google to see whether anyone had similar issues.

- **Engage with the community**: Join the Discord server (`discord.gg/autogpt`) and feel free to ask there if you have any problems. But please be precise with your requests; you are more likely to get help if you provide more context, as messages such as *"My autogpt broke, fix it"* are not very helpful and require much more time than saying what you have already done in detail.

- **Read the documentation**: The whole team is working hard on keeping it up to date and as easy to get into as possible; you will find most information there.

- **Follow steps as they are written**: I personally stopped responding to users who cannot get my Telegram plugin to run and copy/paste their logs, which state they didn't install the correct dependencies. If a README or document tells you to install a certain version of a dependency, please do so as it can become very exhausting to explain to each user individually that they should not assume.

- **Provide feedback and contribute**: If you come across any issues or have suggestions for improving Auto-GPT, don't hesitate to provide feedback or contribute to the project. Open source projects such as Auto-GPT thrive on community involvement.

- **Be patient**: Auto-GPT is an open source project, so the contributors and developers do not get paid and volunteer to work on it for fun. Do not be the person to ruin the fun, and have fun using Auto-GPT!

Download the example code files

You can download the example code files for this book from GitHub at `https://github.com/PacktPublishing/Unlocking-the-Power-of-Auto-GPT-and-Its-Plugins`. If there's an update to the code, it will be updated in the GitHub repository.

We also have other code bundles from our rich catalog of books and videos available at `https://github.com/PacktPublishing/`. Check them out!

Conventions used

There are a number of text conventions used throughout this book.

`Code in text`: Indicates code words in text, database table names, folder names, filenames, file extensions, pathnames, dummy URLs, user input, and Twitter handles. Here is an example: " To verify the installation, run the aforementioned `python --version` command in the command prompt."

A block of code is set as follows:

```
def can_handle_user_input(self, user_input: str) -> bool:
    return True

def user_input(self, user_input: str) -> str:
    return self.telegram_utils.ask_user(prompt=user_input)
```

Any command-line input or output is written as follows:

```
python -m autogpt
```

Bold: Indicates a new term, an important word, or words that you see onscreen. For instance, words in menus or dialog boxes appear in **bold**. Here is an example: "These are then found in the official Auto-GPT Discord server in the **#plugins** channel "

> **Tips or important notes**
> Appear like this.

Get in touch

Feedback from our readers is always welcome.

General feedback: If you have questions about any aspect of this book, email us at customercare@packtpub.com and mention the book title in the subject of your message.

Errata: Although we have taken every care to ensure the accuracy of our content, mistakes do happen. If you have found a mistake in this book, we would be grateful if you would report this to us. Please visit www.packtpub.com/support/errata and fill in the form.

Piracy: If you come across any illegal copies of our works in any form on the internet, we would be grateful if you would provide us with the location address or website name. Please contact us at copyright@packt.com with a link to the material.

If you are interested in becoming an author: If there is a topic that you have expertise in and you are interested in either writing or contributing to a book, please visit authors.packtpub.com.

Share Your Thoughts

Once you've read *Unlocking the Power of Auto-GPT and Its Plugins*, we'd love to hear your thoughts! Scan the QR code below to go straight to the Amazon review page for this book and share your feedback.

https://packt.link/r/1805128280

Your review is important to us and the tech community and will help us make sure we're delivering excellent quality content.

Download a free PDF copy of this book

Thanks for purchasing this book!

Do you like to read on the go but are unable to carry your print books everywhere?

Is your eBook purchase not compatible with the device of your choice?

Don't worry, now with every Packt book you get a DRM-free PDF version of that book at no cost.

Read anywhere, any place, on any device. Search, copy, and paste code from your favorite technical books directly into your application.

The perks don't stop there, you can get exclusive access to discounts, newsletters, and great free content in your inbox daily

Follow these simple steps to get the benefits:

1. Scan the QR code or visit the link below

https://packt.link/free-ebook/9781805128281

2. Submit your proof of purchase
3. That's it! We'll send your free PDF and other benefits to your email directly

1
Introducing Auto-GPT

In the *Preface*, I wrote about what Auto-GPT is and where it came from, but I was asking myself, "*Why would anyone read this book?*"

I mean, it is what it is – an automated form of **artificial intelligence** (**AI**) that may or may not help you do some tasks or be a fun toy that can be very spooky sometimes, right?

I want you to have a clear understanding of what you can or cannot do with it.

Of course, the more creative you get, the more it can do, but sometimes the boundaries appear to be more or less random. For example, let's say you just built a house-building robot that for no apparent reason refuses to make the front door blue, even though you really want a blue door; it keeps going off-topic or even starts explaining what doors are.

Auto-GPT can be very frustrating when it comes to these limitations as they come from a combination of OpenAI's restrictions (which they give in their GPT model) and the humans who write and edit Auto-GPT (along with you – the user who gives it instructions). What first appears to be a clear instruction can result in a very different outcome just by changing one single character.

For me, this is what makes it fascinating – you can always expect it to behave like a living being that can randomly choose to do otherwise and have its own mind.

> **Note**
>
> Always keep in mind that this is a fast-moving project, so code can and will be changed until this book is released. It may also be the case that you bought this book much later and Auto-GPT is completely different. Most of the content in this book focuses on version 0.4.1, but changes have been made and considered regarding version 0.5.0 as well.
>
> For example, once I finished the draft of this book, the "Forge" (an idea we had at a team meeting) had already been implemented. This was an experiment that allowed other developers to build their own Auto-GPT variation.

The Auto-GPT project is a framework that contains Auto-GPT, which we'll be working with in this book, and can start other agents made by other developers. Those agents are in the repositories of the programmers who added them, so we won't dive into them here.

In this chapter, we aim to introduce you to Auto-GPT, including its history and development, as well as LangChain. This chapter will help you understand what Auto-GPT is, its significance, and how it has evolved. By the end of this chapter, you will have a solid foundation to build upon as we explore more advanced topics in the subsequent chapters.

We will cover the following main topics in this chapter:

- Overview of Auto-GPT

- History and development of Auto-GPT

- Introduction to LangChain

Overview of Auto-GPT

Auto-GPT is more or less a category of what it already describes:

"An automated generative pretrained transformer"

This means it automates GPT or ChatGPT. However, in this book, the main focus is on Auto-GPT by name. If you haven't heard of it and just grabbed this book out of curiosity, then you're in the right place!

Auto-GPT started as an experimental self-prompting AI application that is an attempt to create an autonomous system capable of creating "agents" to perform various specialized tasks to achieve larger objectives with minimal human input. It is based on OpenAI's GPT and was developed by *Toran Bruce Richards*, who is better known by his GitHub handle *Significant Gravitas*.

Now, how does Auto-GPT think? Auto-GPT creates prompts that are fed to **large language models (LLMs)** and allows AI models to generate original content and execute command actions such as browsing, coding, and more. It represents a significant step forward in the development of autonomous AI, making it the fastest-growing open source project in GitHub's history (at the time of writing).

Auto-GPT strings together multiple instances of OpenAI's language model – **GPT** – and by doing so creates so-called "agents" that are tasked with simplified tasks. These agents work together to accomplish complex goals, such as writing a blog, with minimal human intervention.

Now, let's talk about how it rose to fame.

From an experiment to one of the fastest-growing GitHub projects

Auto-GPT was initially named **Entrepreneur-GPT** and was released on March 16, 2023. The initial goal of the project was to give GPT-4 autonomy to see if it could thrive in the business world and test its capability to make real-world decisions.

For some time, the development of Auto-GPT remained mostly unnoticed until late March 2023. However, on March 30, 2023, Significant Gravitas tweeted about the latest demo of Auto-GPT and posted a demo video, which began to gain traction. The real surge in interest came on April 2, 2023, when computer scientist Andrej Karpathy quoted one of Significant Gravitas' tweets, saying that the next frontier of prompt engineering was Auto-GPT.

This tweet went viral, and Auto-GPT became a subject of discussion on social media. One of the agents that was created by Auto-GPT, known as **ChaosGPT**, became particularly famous when it was humorously assigned the task of "destroying humanity," which contributed to the viral nature of Auto-GPT (`https://decrypt.co/126122/meet-chaos-gpt-ai-tool-destroy-humanity`).

Of course, we don't want to destroy humanity; for a reference on what Entrepreneur-GPT can do, take a look at the old logs of Entrepreneur-GPT here:

`https://github.com/Significant-Gravitas/Auto-GPT/blob/c6f61db06cde7bd766e521bf7df1dc0c2285ef73/`.

The more creative you are with your prompts and configuration, the more creative Auto-GPT will be. This will be covered in *Chapter 2* when we run our first Auto-GPT instance together.

LLMs – the core of AI

Although Auto-GPT can be used with other LLMs, it best leverages the power of GPT-4, a state-of-the-art language model by OpenAI.

It offers a huge advantage for users who don't own a graphics card that can hold models such as GPT-4 equivalents. Although there are many 7-B and 13-B LLMs (**B** stands for **billion parameters**) that do compete with ChatGPT, they cannot hold enough context in each prompt to be useful or are just not stable enough.

At the time of writing, GPT-4 and GPT-3.5-turbo are both used with Auto-GPT by default. Depending on the complexity of the situation, Auto-GPT differs between two types of models:

- Smart model
- Fast model

When does Auto-GPT use GPT-3.5-turbo and not GPT-4 all the time?

When Auto-GPT goes through its thought process, it uses the *fast model*. For example, as Auto-GPT loops through its thoughts, it uses the configured fast model, but when it summarizes the content of a website or writes code, it will decide to use the smart model.

The default for the fast model is GPT-3.5-turbo. Although it isn't as precise as GPT-4, its response time is much better, leading to a more fluent response time; GPT-4 can seem stuck if it thinks for too long.

OpenAI has also added new functionalities to assist applications such as Auto-GPT. One of them is the *ability to call functions*. Before this new feature, Auto-GPT had to explain to GPT what a command is and how to formulate it correctly in text. This resulted in many errors as GPT sometimes decides to change the syntax of the output that's expected. This was a huge step forward as this feature now reduces the complexity of how commands are communicated and executed. This empowers GPT to better understand what the context of each task is.

So, why don't we use an LLM directly? Because LLMs are only responsive:

- They cannot fulfill any tasks
- Their knowledge is fixed, and they cannot update it themselves
- They don't remember anything; only frameworks that run them can do it

How does Auto-GPT make use of LLMs?

Auto-GPT is structured in a way that it takes in an initial prompt from the user via the terminal:

```
Welcome to Auto-GPT!  run with '--help' for more information.
Create an AI-Assistant:  input '--manual' to enter manual mode.
   Asking user via keyboard...
I want Auto-GPT to: █
```

Figure 1.1 – Letting Auto-GPT define its role

Here, you can either define a main task or enter --manual to then answer questions, as shown here:

```
I want Auto-GPT to: --manual
Manual Mode Selected
Create an AI-Assistant:  Enter the name of your AI and its role below. Entering nothing will load defaults.
Name your AI:  For example, 'Entrepreneur-GPT'
   Asking user via keyboard...
AI Name:
Entrepreneur-GPT here!  I am at your service.
Describe your AI's role:  For example, 'an AI designed to autonomously develop and run businesses with the sole goal of increasing your net worth.'
   Asking user via keyboard...
Entrepreneur-GPT is:
Enter up to 5 goals for your AI:  For example: Increase net worth, Grow Twitter Account, Develop and manage multiple businesses autonomously'
   Enter nothing to load defaults, enter nothing when finished.
   Asking user via keyboard...
Goal 1: Help people get into Auto-GPT by writing a book
   Asking user via keyboard...
Goal 2: make sure to write all research into local files and keep them short, max 2000 charactars per file
   Asking user via keyboard...
Goal 3: if a file gets too long, chunk it by using the suffix _partX.txt
   Asking user via keyboard...
Goal 4: create an index.txt file for each of those files, with 2-3 sentences to describe it.
   Asking user via keyboard...
Goal 5: █
```

Figure 1.2 – Setting Auto-GPT's main goals

The main prompt is then saved as an `ai_settings.yaml` file that may look like this:

```
ai_goals:
- Conduct a thorough analysis of the current state of the book
  and identify areas for improvement.
- Develop a comprehensive plan for creating task lists that will help
you structure research, a detailed outline per chapter and individual
parts.
- Be sure to ask the user for feedback and improvements.
- Continuously assess the current state of the work and use the speak
property to give the user positive affirmations.
ai_name: AuthorGPT
ai_role: An AI-powered author and researcher specializing in creating
comprehensive, well-structured, and engaging content on Auto-GPT and
its plugins, while maintaining an open line of communication with the
user for feedback and guidance.
api_budget: 120.0
```

Let's look at some of the AI components in the preceding file:

- First, we have `ai_goals`, which specifies the main tasks that Auto-GPT must undertake. It will use those to decide which individual steps to take. Each iteration will decide to follow one of the goals.

- Then, we have `ai_name`, which is also taken as a reference and defines parts of the behavior or character of the bot. This means that if you call it *AuthorGPT*, it will play the role of a GPT-based author, while if you call it *Author*, it will try to behave like a person. It is generally hard to tell how it will behave because GPT mostly decides what it puts out on its own.

- Finally, we have `ai_role`, which can be viewed as a more detailed role description. However, in my experience, it only nudges the thoughts slightly. Goals are more potent here.

Once this is done, it summarizes what it's going to do and starts thinking correctly:

```
Goal 4: create an index.txt file for each of those files, with 2-3 sentences to describe it.
  Asking user via keyboard...
Goal 5:
Enter your budget for API calls:  For example: $1.50
  Enter nothing to let the AI run without monetary limit
  Asking user via keyboard...
Budget: $2
Entrepreneur-GPT  has been created with the following details:
Name:  Entrepreneur-GPT
Role:  an AI designed to autonomously develop and run businesses with the
Goals:
-  Help people get into Auto-GPT by writing a book
-  make sure to write all research into local files and keep them short, max 2000 charactars per file
-  if a file gets too long, chunk it by using the suffix _partX.txt
-  create an index.txt file for each of those files, with 2-3 sentences to describe it.
Using memory of type:  JSONFileMemory
Using Browser:  chrome
read more here: https://docs.agpt.co/setup/#getting-an-api-key
 Thinking...
```

Figure 1.3 – Example of Auto-GPT's thought process

Thinking generally means that it is sending a chat completion request to the LLM.

This process can be slow – the more tokens that are used, the more processing that's needed. In the *Understanding tokens in LLMs* section, we will take a look at what this means.

Once Auto-GPT has started "thinking," it initiates a sequence of AI "conversations." During these conversations, it forms a query, sends it to the LLM, and then processes the response. This process repeats until it finds a satisfactory solution or reaches the end of its thinking time.

This entire process produces thoughts. These fall into the following categories:

- Reasoning

- Planning

- Criticism

- Speak

- Command

These individual thoughts are then displayed in the terminal and the user is asked whether they want to approve the command or not – it's that simple.

Of course, a lot more goes on here, including a prompt being built to create that response.

Simply put, Auto-GPT passes the name, role, goals, and some background information. You can see an example here: `https://github.com/PacktPublishing/Unlocking-the-Power-of-Auto-GPT-and-Its-Plugins/blob/main/Auto-GPT_thoughts_example.md`.

Auto-GPT's thought process – understanding the one-shot action

Let's understand the thought process behind this one-shot action:

- **Overview of the thought process**: Auto-GPT operates on a one-shot action basis. This approach involves processing each data block that's sent to OpenAI as a single chat completion action. The outcome of this process is that a response text from GPT is generated that's crafted based on a specified structure.

- **Structure and task definition for GPT**: The structure that's provided to GPT encompasses both the task at hand and the format for the response. This dual-component structure ensures that GPT's responses are not only relevant but also adhere to the expected conversational format.

- **Role assignment in Auto-GPT**: There are two role assignments here:

  - **System role**: The "system" role is crucial in providing context. It functions as a vessel for information delivery and maintains the historical thread of the conversation with the LLM.

- **User role**: Toward the end of the process, a "user" role is assigned. This role is pivotal in guiding GPT to determine the subsequent command to execute. It adheres to a predefined format, ensuring consistency in interactions.

- **Command options and decision-making**: GPT is equipped with various command options, including the following:

  - Ask the user (`ask_user`)

  - Sending messages (`send_message`)

  - Browsing (`browse`)

  - Executing code (`execute_code`)

In some instances, Auto-GPT may opt not to select any command. This typically occurs in situations of confusion, such as when the provided task is unclear or when Auto-GPT completes a task and requires user feedback for further action.

Either way, each response is only one text and just a text that is being autocompleted, meaning the LLM only responds once with such a response.

In the following example, I have the planner plugin activated; more on plugins later:

```
{
"thoughts": {
"text": "I need to start the planning cycle to create a plan for the
book.",
"reasoning": "Starting the planning cycle will help me outline the
steps needed to achieve my goals.",
"plan":
"- run_planning_cycle
- research Auto-GPT and its plugins
- collaborate with user
- create book structure
- write content
- refine content based on feedback",
"criticism": "I should have started the planning cycle earlier to
ensure a smooth start.",
"speak": "I'm going to start the planning cycle to create a plan for
the book."
},
"command": {
"name": "run_planning_cycle",
"args": {}
}
}
```

Each thought property is then displayed to the user and the "speak" output is read aloud if text-to-speech is enabled:

```
"I am going to start the planning cycle to create a plan for the book.
I want to run planning cycle."
```

The user can now respond in one of the following ways:

- y: To accept the execution.

- n: To decline the execution and close Auto-GPT.

- s: To let Auto-GPT re-evaluate its decisions.

- y -n: To tell Auto-GPT to just keep going for the number of steps (for example, enter y -5 to allow it to run on its own for 5 steps). Here, n is always a number.

If the user confirms, the command is executed and the result of that command is added as system content:

```
# Check if there is a result from the command append it to the message
# history
if result is not None:
self.history.add("system", result, "action_result")
```

At this point, you're probably wondering what history is in this context and why self?

Auto-GPT uses agents and the instance of the agent has its own history that acts as a short-term memory. It contains the context of what the previous messages and results were.

The history is trimmed down on every run cycle of the agent to make sure it doesn't reach its token limit.

So, why not directly ask the LLM for a solution? There are several reasons for this:

- While LLMs are incredibly sophisticated, they cannot solve complex, multi-step problems in a single query. Instead, they need to be asked a series of interconnected questions that guide them toward a final solution. This is where Auto-GPT shines – it can strategically ask these questions and digest the responses.

- LLMs can't maintain their context. They don't remember previous queries or answers, which means they cannot build on past knowledge to answer future questions. Auto-GPT compensates for this by maintaining a history of the conversation, allowing it to understand the context of previous queries and responses and use that information to craft new queries.

- While LLMs are powerful tools for generating human-like text, they cannot take initiative. They respond to prompts but don't actively seek out new tasks or knowledge. Auto-GPT, on the other hand, is designed to be more proactive. It not only responds to the tasks that have been assigned to it but also proactively explores diverse ways to accomplish those tasks, making it a true autonomous agent.

Before we delve deeper into how Auto-GPT utilizes LLMs, it's important to understand a key component of how these models process information: **tokens**.

Understanding tokens in LLMs

Tokens are the fundamental building blocks in LLMs such as GPT-3 and GPT-4. They are pieces of knowledge that vary in proximity to each other based on the given context. A token can represent a word, a symbol, or even fragments of words.

Tokenization in language processing

When training LLMs, text data is broken down into smaller units, or tokens. For instance, the sentence "ChatGPT is great!" would be divided into tokens such as `["ChatGPT", "is", "great", "!"]`. The nature of a token can differ significantly across languages and coding paradigms:

- In English, a token typically signifies a word or part of a word

- In other languages, a token may represent a syllable or a character

- In programming languages, tokens can include keywords, operators, or variables

Let's look at some examples of tokenization:

- **Natural language**: The sentence "ChatGPT is great!" tokenizes into `["ChatGPT", "is", "great", "!"]`.

- **Programming language**: A Python code line such as `print("Hello, World!")` is tokenized as `["print", "(", " ", "Hello", "," , " ", "World", "!"", ")"]`.

Balancing detail and computational resources

Tokenization strategies aim to balance detail and computational efficiency. More tokens provide greater detail but require more resources for processing. This balance is crucial for the model's ability to understand and generate text at a granular level.

Token limits in LLMs

The token limit signifies the maximum number of tokens that a model such as GPT-3 or GPT-4 can handle in a single interaction. This limit is in place due to the computational resources needed to process large numbers of tokens.

The token limit also influences the model's "attention" capability – its ability to prioritize different parts of the input during output generation.

Implications of token limits

A model with a token limit may not fully process inputs that exceed this limit. For example, with a 20-token limit, a 30-token text would need to be broken into smaller segments for the model to process them effectively.

In programming, tokenization aids in understanding code structure and syntax, which is vital for tasks such as code generation or interpretation.

In summary, tokenization is a critical component in **natural language processing** (**NLP**), enabling LLMs to interpret and generate text in a meaningful and contextually accurate manner.

For instance, if you're using the model to generate Python code and you input `["print", "("]` as a token, you'd expect the model to generate tokens that form a valid argument to the print function – for example, `[""Hello, World!"", ")"]`.

In the following chapters, we will delve deeper into how Auto-GPT works, its capabilities, and how you can use it to solve complex problems or automate tasks. We will also cover its plugins, which extend its functionality and allow it to interact with external systems so that it can order a pizza, for instance.

In a nutshell, Auto-GPT is like a very smart, very persistent assistant that leverages the power of the most advanced AI to accomplish the goals you set for it. Whether you're an AI researcher, a developer, or simply someone who is fascinated by the potential of AI, I hope this book will provide you with the knowledge and inspiration you need to make the most of Auto-GPT.

At the time of writing (June 1, 2023), Auto-GPT can give you feedback not only through the terminal. There are a variety of text-to-speech engines that are currently built into Auto-GPT. Depending on what you prefer, you can either use the default, which is Google's text-to-speech option, ElevenLabs, macOS' `say` command (a low-quality Siri voice pack), or Silero TTS.

When it comes to plugins, Auto-GPT becomes even more powerful. Currently, there is an official repository for plugins that contains a list of awesome plugins such as Planner Plugin, Discord, Telegram, Text Generation for local or different LLMs, and more.

This modularity makes Auto-GPT the most exciting thing I've ever laid my hands on.

Launching and advancing Auto-GPT – a story of innovation and community

Auto-GPT's development began with a bold vision to make the sophisticated technology of GPT-4 accessible and user-friendly. This initiative marked the start of an ongoing journey, with the project continually evolving through the integration of new features and improvements. At its core, Auto-GPT is a collaborative effort, continuously shaped by the input of a dedicated community of developers and researchers.

The genesis of Auto-GPT can be traced back to the discovery of GPT-4's potential for autonomous task completion. This breakthrough was the catalyst for creating a platform that could fully utilize GPT-4's capabilities, offering users extensive control and customization options.

The project gained initial popularity with an early version known as *Entrepreneur-GPT*, a key milestone that showcased Auto-GPT's capabilities at the time. This phase of the project (documented here:

`https://github.com/PacktPublishing/Unlocking-the-Power-of-Auto-GPT-and-Its-Plugins/blob/main/Entrepreneur-GPT.md`) indicates the differences in prompts and functionalities compared to later stages. A review of the git history reveals Auto-GPT's early abilities, including online research and using a local database for long-term memory.

The ascent of Auto-GPT was swift, attracting contributors – including myself – early in its development. My experience with this open source project was transformative, offering an addictive blend of passion and excitement for innovation. The dedication of the contributors brought a sense of pride, especially when you can see your work recognized by a wider audience, including popular YouTubers.

As an open source project, Auto-GPT thrived on voluntary contributions, leading to the formation of a team that significantly enhanced its structure. This team played a crucial role in managing incoming pull requests and guiding the development paths, thereby continually improving Auto-GPT's core.

Despite its growing popularity, each new release of Auto-GPT brought enhanced power and functionality. These releases are stable versions that are meticulously tested by the community to ensure they are bug-free and ready for public use.

A critical component of Auto-GPT's evolution is its plugins. These play a major role in the customization of the platform, allowing users to tailor it to their specific needs. Future discussions will delve deeper into these plugins and will explore their installation, usage, and impact on enhancing Auto-GPT's capabilities. This exploration is vital as most customization happens through plugins unless significant contributions are made directly to the core platform through pull requests.

Introduction to LangChain

Although *LangChain* itself is not part of Auto-GPT, it is a crucial component of Auto-GPT's development as it focuses on the process using control. This is in contrast to Auto-GPT's emphasis on results without control.

LangChain is a powerful tool that enables users to build implementations of their own Auto-GPT using LLM primitives. It allows for explicit reasoning and the potential for Auto-GPT to become an autonomous agent.

With multiple alternatives of Auto-GPT arising, LangChain has become a part of many of them. One such example is AgentGPT.

LangChain's unique approach to language processing and control makes it an essential part of AgentGPT's functionality. By combining the strengths of LangChain and Auto-GPT, users can create powerful, customized solutions that leverage the full potential of GPT.

The intersection of LangChain and Auto-GPT

LangChain and Auto-GPT may have different areas of focus, but their shared goal of enhancing the capabilities of LLMs creates a natural synergy between them. LangChain's ability to provide a structured, controllable process pairs well with Auto-GPT's focus on autonomous task completion. Together, they provide an integrated solution that both controls the method and achieves the goal, striking a balance between the process and the result.

LangChain enables the explicit reasoning potential within Auto-GPT. It provides a pathway to transition the model from being a tool for human-directed tasks to a self-governing agent capable of making informed, reasoned decisions.

In addition, LangChain's control over language processing enhances Auto-GPT's ability to communicate user-friendly information in JSON format, making it an even more accessible platform for users. By optimizing language processing and control, LangChain significantly improves Auto-GPT's interaction with users.

You can read more about it: `https://docs.langchain.com/docs/`.

Summary

In this chapter, we embarked on the exciting journey of exploring Auto-GPT, an innovative AI application that leverages the power of GPT-4 to autonomously solve tasks and operate in a browser environment. We delved into the history of Auto-GPT, understanding how it evolved from an ambitious experiment to a powerful tool that's transforming the way we interact with AI.

We also explored the concept of tokens, which play a crucial role in how LLMs such as GPT-4 process information. Understanding this fundamental concept will help us better comprehend how Auto-GPT interacts with LLMs to generate meaningful and contextually relevant responses.

Furthermore, we touched on the role of LangChain, a tool that complements Auto-GPT by providing structured control over language processing. The intersection of LangChain and Auto-GPT creates a powerful synergy, enhancing the capabilities of Auto-GPT and paving the way for more advanced AI applications.

As we move forward, we will dive deeper into the workings of Auto-GPT, exploring its plugins, installation process, and how to craft effective prompts. We will also delve into more advanced topics, such as integrating your own LLM with Auto-GPT, setting up Docker, and safely and effectively using continuous mode.

Whether you're an AI enthusiast, a developer, or simply someone curious about the potential of AI, this journey promises to be a fascinating one. So, buckle up, and let's continue to unravel the immense potential of Auto-GPT together!

2

From Installation to Your First AI-Generated Text

Now that we have finished the first chapter, let's discuss the must-have requirements of **Auto-GPT**.

At this point, before we start, whether you choose to register for an **application programming interface** (**API**) account or not at OpenAI is up to you. I first recommend trying to install and start Auto-GPT before registering, in case Auto-GPT only works in Docker (which could happen as it keeps changing); it may or may not be possible for you to run Auto-GPT. However, let's begin by setting up Auto-GPT without the account. Otherwise, you will have an OpenAI account but there is no need for it.

In this chapter, we will guide you through preparing your machine to run Auto-GPT, the installation process, and your first steps with Auto-GPT.

We'll cover the fundamental concepts, installation, and setup instructions for Auto-GPT. We will conclude by explaining how to execute your first AI-automated task using Auto-GPT.

The team at Auto-GPT (including me) is working hard on making Auto-GPT as accessible as possible.

Recently a new tool has been made by one of the maintainers called **Auto-GPT Wizard**. If you struggle with setting up Auto-GPT at any point, this tool is meant to automate the installation and make it easier for newbies to get into Auto-GPT.

You can find the tool at `https://github.com/Significant-Gravitas/AutoGPT_Wizard`.

In this chapter, we will learn about the following topics:

- System requirements and prerequisites
- Installing and setting up Auto-GPT

- Going through the basic concepts and terminology
- First run

Here are some system requirements and prerequisites:

- Install VS Code.
- Install Python.
- Install Poetry.

Installing VS Code

I strongly recommend installing VS Code for usability or using any other IDE you see fit for Python. Working as a triage catalyst (reviewer, support, and contributor role at Auto-GPT), I have encountered many people who got stuck because they used their text editor or even Microsoft Word.

Using advanced text editors configured properly might be adequate for basic scripting or editing configuration files, as they can be configured to avoid issues with text encoding and incorrect file extensions. However, IDEs such as VS Code offer more robust tools and integrations for a seamless development experience, especially when dealing with complex projects such as Auto-GPT; but we will have to edit JSON files, a `.env` file, and sometimes markdown (`.md`) files. Editing those with anything else than an IDE will probably result in the wrong file extension being added (for example, `.env` and `settings.json` could become `.env.txt` or `settings.json.docx`, which do not work).

As a common tool to be used by many developers and it being free to use, we will focus on VS Code.

To not drift off the topic of why else you could use VS Code, Microsoft wrote a very good article on why VS Code is worth using. Of course, you can also use other IDEs. The main reason I recommend VS Code is that it is open source and free to use and also used by most Auto-GPT contributors, making it very easy to work with Auto-GPT and some integrated project settings for VS Code.

Installing Python 3.10

If you want to run Auto-GPT directly without Docker, you may need to install Python 3.10 or enable it as the terminal's `python` and `python3` alias, to make sure that Auto-GPT doesn't accidentally call a different Python version.

Auto-GPT is developed in Python and it specifically requires Python version 3.10.x. The x in 3.10.x represents any sub-version (for example, 3.10.0, 3.10.6), and the software is compatible with these sub-versions.

While Auto-GPT is lightweight in terms of file size, it can be resource-intensive depending on the options and plugins you enable. Consequently, it's essential to have a compatible and optimized environment to ensure the smooth operation of Auto-GPT and the plugins you may choose to use, as those are all written for Python 3.10 and those expected modules that are also for 3.10. In addition to installing Python 3.10, it is recommended to use virtual environments for Auto-GPT development. Virtual environments allow you to manage dependencies and Python versions on a project-by-project basis, ensuring that Auto-GPT runs in an isolated and controlled setting without affecting other Python projects you may be working on. This is crucial for maintaining compatibility and avoiding conflicts between different project requirements.

Why choose Python 3.10?

Python 3.10 introduces several features and optimizations that are beneficial for running Auto-GPT. One such feature is the improved syntax for type hinting. In Python 3.10, you can use the pipe symbol, |, as a more concise way of indicating that a variable can be of multiple types. This is known as the **type union operator**:

```
File "L:\00000000ai\Auto-GPT\autogpt\llm\providers\openai.py", line
95, in <module>
    OPEN_AI_MODELS: dict[str, ChatModelInfo | EmbeddingModelInfo |
TextModelInfo] = {
TypeError: unsupported operand type(s) for |: 'type' and 'type'
```

In this example error message, Auto-GPT is attempting to use this new type union syntax, which is not supported in Python versions earlier than 3.10. This is why using Python 3.9 results in a syntax error, as it cannot parse the new syntax.

Additionally, Python 3.10 brings performance improvements, better error messages, and new features that can be advantageous for complex applications such as Auto-GPT.

Therefore, to avoid compatibility issues and take advantage of the new features and optimizations, it is crucial to install and set up Python 3.10 correctly before running Auto-GPT.

Prerequisites for Python Installation

Before installing Python 3.10, it's important to ensure that your system meets the necessary prerequisites. These prerequisites include the following:

- **Sufficient disk space**: Make sure your system has an adequate amount of free disk space to accommodate the Python installation and any additional packages or libraries you may install in the future.

- **Checking for existing Python installations**: If you already have a previous version of Python installed on your system, it's recommended to check for any potential conflicts or compatibility issues that may arise with Python 3.10. You can do this by running the appropriate version-specific commands or using the Python version management tool for your **operating system (OS)**.

By ensuring that your system meets these prerequisites, you can proceed with confidence to install Python 3.10 and set up Auto-GPT successfully.

Installing Python 3.10

Auto-GPT is mostly based on Python 3.10 packages; if you try to run it with 3.9 for example, you will only receive a few exceptions and you will not be able to execute Auto-GPT.

Running Python 3.10 requires a system that can support this version of the programming language. Here are the system requirements and installation instructions for each OS:

- For Windows, use the documentation at https://www.digitalocean.com/community/tutorials/install-python-windows-10.

  To verify the installation, run the following command in the command prompt:

  ```
  python --version
  ```

  The system should return Python 3.10.x.

- To install Python 3.10 on macOS, use the documentation at https://www.digitalocean.com/community/tutorials/how-to-install-python-3-and-set-up-a-local-programming-environment-on-macos.

  To verify the installation, run the aforementioned python --version command in the command prompt.

- For installing Python 3.10 on Linux (Ubuntu/Debian), you may have to do a bit of research depending on what flavor of Linux you are using. But, as they say, with great power comes great responsibility; you may have to research how to enable Python 3.10 on your machine.

- For Ubuntu and Debian, here is the documentation on how to install 3.10: https://www.linuxcapable.com/how-to-install-python-3-10-on-ubuntu-linux/.

  To verify if the installation was done successfully, run the following command:

  ```
  python3.10 -version
  ```

  The system should return Python 3.10.x.

> **Note**
>
> The exact commands and steps might vary slightly based on the specific version of each OS. Always refer to the official Python documentation or your OS's documentation for the most correct and up-to-date information.

Installing Poetry

A new dependency was recently added, which is a bit tricky to install.

Documentation on how to install it can be found here: `https://python-poetry.org/docs/#installing-with-pipx`.

If you struggle to set it up (on Windows, for example), you may also try the Wizard script here: `https://github.com/Significant-Gravitas/AutoGPT_Wizard`.

Additional requirements that may come up

Please check the official documentation at `https://docs.agpt.co/autogpt/setup/` to make sure you are not missing anything.

In addition to having a compatible version of Python and poetry installed on your system, it is also essential to ensure that your hardware meets specific minimum requirements for running Auto-GPT effectively:

- **Processor (CPU)**: A modern multi-core processor (Intel Core i5/i7 or AMD Ryzen) is recommended for optimal performance when using Auto-GPT

- **Memory (RAM)**: A minimum of 8 GB RAM is recommended; however, having more memory available will improve performance when working with large datasets or complex tasks

- **Storage**: Ensure sufficient free disk space on your computer's primary storage drive (HDD/SSD) – at least several gigabytes – as Auto-GPT may generate temporary files during operation and require additional space for storing generated output files

- **Internet connection**: A stable internet connection with reasonable bandwidth is necessary since Auto-GPT communicates with OpenAI's API to access GPT models and generate text

- **GPU support (optional)**: While not strictly required, having a compatible NVIDIA or AMD GPU can significantly improve the performance of certain tasks, such as using text-to-speech engines such as Silero **Text-to-Speech (TTS)**

By ensuring that your system meets these requirements and prerequisites, you will be well prepared to install and use Auto-GPT effectively.

In the next sections, we will guide you through the installation process for Auto-GPT on various OSs and provide an overview of basic concepts and terminology related to Auto-GPT and its underlying technology.

Remember that while these system requirements and prerequisites are designed to provide a smooth experience when using Auto-GPT, individual needs may vary depending on the specific tasks you plan to perform with the tool. For example, if you intend to use Auto-GPT for large-scale text generation or complex **natural language processing (NLP)** tasks, you might benefit from having a more powerful CPU or additional memory available.

In any case, it is always a good idea to monitor your system's performance while using Auto-GPT and adjust your hardware configuration as needed. This will help ensure that you can make the most of this powerful AI-driven text generation tool without meeting performance bottlenecks or other issues.

Before installing and setting up Auto-GPT on your system, do the following:

1. Ensure that your OS (macOS, Linux/Ubuntu/Debian, Windows) meets the minimum requirements for running Python 3.10.

2. Install Python 3.10.x, following the instructions provided for each OS.

3. Verify that Python 3.10.x has been installed correctly by checking its version in Terminal (macOS/Linux) or Command Prompt (Windows).

4. Make sure your hardware meets minimum requirements such as processor (CPU), memory (RAM), storage space availability, and internet connection stability/bandwidth.

5. This step is optional. Consider GPU support if planning to use resource-intensive features such as text-to-speech engines or a local LLM such as Vicuna or LLAMA (this is an interesting topic, as most GPUs cannot handle an LLM that's usable with Auto-GPT).

By following these guidelines carefully and ensuring that your system meets all requirements and prerequisites before installing Auto-GPT, you will be well prepared for a successful installation process and an enjoyable experience using this powerful AI-driven text generation tool.

In our next sections, we will guide you through every step of getting started with this incredible software – from installation procedures tailored specifically for each OS to understanding the fundamental concepts and terminology that underpin Auto-GPT's functionality.

Installing and setting up Auto-GPT

Here are the steps to install Auto-GPT:

1. Depending on your experience, you may want to either use Git directly and clone the repository from `https://github.com/Significant-Gravitas/Auto-GPT.git`. Or, if you are less experienced with the Terminal, you may go to `https://github.com/Significant-Gravitas/Auto-GPT`.

2. On the top right, click on **Code**, then **Local**, download it as a `.zip` file, and save it anywhere you want Auto-GPT's folder to be. Then, simply unpack the `.zip` file.

3. If you want to be 100% sure that you are on the most stable version, go to `https://github.com/Significant-Gravitas/Auto-GPT/releases/latest`.

4. Pick the latest release (in our case, 0.4.1) download the `.zip` file in the **Assets** section of that post, and unzip it.

5. The latest version that I used was release v0.4.7; anything above that version may be restructured, for example, 0.5.0 already has the Auto-GPT folder inside `Auto-GPT/autogpts/autogpt`. For closer inspection, read the updated `README` and documentation inside the repository itself to see which version you are looking at.

Installing Auto-GPT

For a fast-growing project, the installation of Auto-GPT may differ, so if you have any trouble with the following guide, try to check `https://docs.agpt.co/` if anything has changed.

As Auto-GPT on its own comes with a variety of Python dependencies, you may now want to navigate to your Auto-GPT folder in a Terminal.

Using Auto-GPT with Docker, do the following:

1. Some developers use Dockerfile directly, but I (as a Docker newbie) recommend using `docker-compose.yml`, which some folks have added.

2. Make sure you have Docker installed (go back to *Installing Docker* in the previous chapter).

3. Simply navigate into the Auto-GPT folder and run the following commands:

```
docker-compose build auto-gpt
docker-compose run -rm auto-gpt -gpt3only
```

> **Note**
>
> I am giving `-gpt3only` as an example only to make sure we don't spend any money yet, as I assume you have just created your OpenAI account, which grants a free $5 starting bonus.

Using Docker to pull the Auto-GPT image

Here, let's ensure you have Docker installed on your system. If you are not sure, you can jump to *Chapter 6*, where I cover how to set up Docker on your machine and give you some extra tips on using Docker with Auto-GPT.

If you have Docker installed, do the following steps:

1. Create a project directory for Auto-GPT:

```
mkdir Auto-GPT
cd Auto-GPT
```

2. In the project directory, create a file called `docker-compose.yml` with the specified contents provided in the documentation.

3. Create the necessary configuration files. You can find templates in the repository.

4. Pull the latest image from Docker Hub:

```
docker pull significantgravitas/auto-gpt
```

5. Run with Docker according to the instructions given in the documentation.

Cloning Auto-GPT using Git

Assuming you have Git installed on your system (doesn't come natively on Windows for example), we will cover the process of cloning Auto-GPT here.

Let's ensure you have Git installed for your OS:

1. We need to first clone the repository with the help of the following command:

```
git clone -b stable https://github.com/Significant-Gravitas/
Auto-GPT.git
```

2. Next, we will navigate to the directory where you downloaded the repository:

```
cd Auto-GPT
python -m pip install -r ./requirements.txt
```

> **Without Git/Docker**
>
> 1. Download the source code (the `.zip` file) from the latest stable release.
>
> 2. Extract the zipped file into a folder.

3. Next, we will navigate to the directory where you downloaded the repository:

```
cd Auto-GPT
python -m pip install -r ./requirements.txt
```

> **Note**
>
> From here, depending on the version you may install, Auto-GPT could be inside the `Auto-GPT/autogpts/autogpt` folder, as the main repository was turned into more of a framework to be used to create other `Auto-GPT` instances. The Auto-GPT project we discuss in this book is inside the folder mentioned previously.

Configuration

Here is how we configuration happens:

1. Find the file named `.env.template` in the main Auto-GPT folder.

2. Create a copy of `.env.template` and rename it `.env`.

3. Open the `.env` file in a text editor. If you have not already, investigate using VS Code, for example, so that you can just open Auto-GPT as a project and edit anything you need to.

4. Find the line that says `OPENAI_API_KEY=`.

5. Enter your unique OpenAI API key after the = symbol without any quotes or spaces.

6. If you use multiple Auto-GPT instances (which can be easily done with just another `auto-gpt` folder; it is best to create multiple API keys), you can make sure you keep an eye on the costs of each instance.

7. Depending on which GPT model you have access to, you must now change the `FAST_LLM_MODEL` and `SMART_LLM_MODEL` attributes the same way we just did with the API key.

8. To find out which models are available to you, go to `https://platform.openai.com/account/rate-limits`.

9. It only lists the ones you can use.

As of writing this chapter, OpenAI has just released a 16 K model of gpt-3.5-turbo-16k. It can carry more tokens/words than GPT-4 can, but I generally feel like the output is still worse than GPT-4's, as Auto-GPT tends to do random tasks that it makes up out of nowhere.

The issue lies in the context process ability, although it can work with more tokens, GPT-4 has far more parameters than it works with and is much more optimized.

The default number of tokens is 4,000 if you set GPT-3.5-Turbo as a model and 8,000 tokens if you set GPT-4 as a model, but I do recommend setting those limits slightly below those.

For example, 7,000 instead of 8,000 gives less room for memory summarization on the `SMART_LLM_MODEL`, while still making sure there are no exceptions where more words or tokens slip through to the Chat Completion prompt.

Auto-GPT has introduced customizing options such as disallowing certain commands or which text-to-speech engine you want to use.

Having speech enabled makes Auto-GPT talk to you via voice. The choice of which TTS engine to use is all yours. I prefer Silero TTS, as it is almost as good as ElevenLabs but it is completely free to use; you just need a computer with a powerful CPU and/or GPU (you can select whether to use CPU or GPU for the TTS model).

As you already may have noticed, Auto-GPT comes with a huge set of terminologies that come from the world of AI and machine learning. We will now cover some of the most frequent ones here.

Basic concepts and terminologies

Before we start using Auto-GPT, let's review some basic concepts and terminologies that will help us understand how it works:

- **Text generation**: Text generation is the task of creating natural language text from given input data or context. For example, given a topic, a genre, or a prompt, text generation can produce a paragraph, an article, a story, or a dialogue that matches the input.

- **Model**: A model is a mathematical representation of a system or a process that can be used to make predictions or decisions. In machine learning, a model is a function that maps an input to an output. For example, a model can take an image as an input and output a label that describes what is in the image.

- **Chain of thought**: This concept is centered on the progressive development and refining of ideas or solutions through the systematic and sequential application of thought processes. In the context of using a tool such as ChatGPT, a "chain of thought" approach would involve feeding the output from one query as the input to the next query, essentially creating a "chain" of evolving responses.

  This method allows for the deep exploration of a topic or problem, as each step in the chain builds upon the previous, potentially leading to more nuanced and sophisticated results. It could be particularly useful in tasks such as developing a complex narrative, iteratively refining a model, or exploring multiple angles of a problem before settling on a solution.

- **Tree of thought**: A strategy to retrieve much better results in text generation, ChatGPT for example, is to instruct it to solve a problem and provide multiple alternatives. This can be achieved by saying "Write four alternatives, assess them, and improve." This simple instruction tells the model to be creative and create four alternatives to an already given solution, evaluate them, and then encourage it to output an improved solution instead of just one answer.

  This results in much more accurate output and can be chained and done multiple times. For instance, I was working on a new neural cell network prototype and asked ChatGPT to help me with the data transformer method that would receive a string (text) and apply it to multiple matrixes. The first result was bad and wasn't even correct Python code, but after three or four iterations of saying "Write 4 alternatives that may improve that code and improve its strategy, assess them, rate them from 1-10, rank them, then improve," this resulted in very clean code and it even gave me improvement ideas on how to make the code more performant after the second iteration, which it wouldn't have done if I just asked it straight away.

- **Forest of thought**: This one builds on top of the principle of the tree of thought but as the name already suggests, you have multiple instances that think like a group of people. A fantastic explanation can be found in this video I watched recently: `https://www.youtube.com/watch?v=y6SVA3aAfco`.

- **Neural network**: A neural network is a type of model that consists of interconnected units called **neurons**. Each neuron can perform a simple computation on its inputs and produce an output. By combining neurons in different layers and configurations, a neural network can learn complex patterns and relationships from data. GPT, for instance, has multiple neural networks running that all have different tasks and consist of multiple layers of neural networks.

- **Deep learning**: Developed by OpenAI, **Generative Pre-Trained Transformer 3 (GPT-3)** stands as a monumental figure in the realm of NLP. This deep learning model, boasting a staggering 175 billion parameters and a vast training dataset of 45 terabytes, is renowned for its text generation capabilities, offering coherence and versatility across a myriad of topics, genres, and styles. Despite anticipation surrounding its successor, GPT-4, which promises enhanced context understanding and logical processing, GPT-3 remains a formidable tool, especially for smaller tasks. Its recent upgrade to process up to 16 K tokens has notably enhanced output quality, although it is advised to avoid overwhelming the model with excessive input data to prevent confusion.

- **GPT-3**: GPT-3 is a deep learning model for NLP that was developed by OpenAI. It is one of the largest and most powerful models for text generation, with 175 billion parameters and 45 terabytes of training data. GPT-3 can generate coherent and diverse text for almost any topic, genre, or style. It is continuously improved by OpenAI, and although the successor GPT-4 may have more capability in terms of context size and logic, it is a faster model and still very useful for small tasks. It can now process 16 K tokens, but I found that this strength is more useful on outputs and not input data. This means the model gets confused quickly when too much information is provided at once.

- **GPT-4**: This is a successor to GPT-3, which is far more powerful for text generation. It has 170 trillion parameters, almost 1,000 times more than GPT-3. This model receives all plugins and a Bing browser functionality, which allows it to research information on its own. OpenAI is being very secretive about some details, as it is yet unknown how it works in detail. Some resources and papers suggest that it works recursively and learns with each input it gets.

- **Auto-GPT**: Auto-GPT is a tool that automates the process of text generation using OpenAI's Chat Completion API, mainly with GPT-4. It allows you to specify your input text and parameters that control the output text, such as length, tone, format, and keywords. Auto-GPT then sends your input text and parameters to the GPT-3 model via the OpenAI API and receives the generated text as a response. Auto-GPT also supplies features to help you edit and improve the generated text, such as suggestions, feedback, and rewriting:

  - **Plugins**: Extensions that can be loaded into Auto-GPT to add more functionality.

  - **Headless browser**: A web browser without a graphical user interface, used for automated tasks.

  - **Workspace**: The directory where Auto-GPT saves files and data.

- **API key**: An API key is a unique identifier used to authenticate a user, developer, or calling program during an API request. This key helps in tracking and controlling how the API is being used, to prevent abuse and ensure security. Essentially, it acts as a password that grants access to specific services or data, facilitating seamless and secure communication between different software components. It is paramount that API keys are kept confidential to prevent unauthorized access and potential misuse.

First run of Auto-GPT on your machine

To run Auto-GPT, you need to use one of the commands, depending on your OS. Use `run.sh` for Linux or macOS, and `run.bat` for Windows. Alternatively, you can just run the following on your console. Navigate into the Auto-GPT folder (not the one inside – I know the folder structure can be misleading sometimes), and execute the following:

```
python -m autogpt
```

You may also execute the "autogpt.bat" or "autogpt.sh" script inside the "autogpts/autogpt" folder.

If you are not sure whether your default Python is Python 3.10, or if the preceding command returns an error, you can check that with the `python -V` command. Should you get anything but Python 3.10, you can run this instead:

```
python3.10 -m autogpt
```

For any OS, you can also use `docker-compose` if you have Docker installed.

You can also pass some arguments to customize your Auto-GPT experience, such as the following:

- `-gpt3only` to use GPT-3.5 instead of GPT-4
- `-speak` to enable text-to-speech output
- `-continuous` to run Auto-GPT without user authorization (not recommended)
- `-debug` to print out debug logs and more
- You can use `-help` to see the full list of arguments

You can also change the Auto-GPT settings in your `.env` file, such as `SMART_LLM_MODEL` to choose the language model, `DISABLED_COMMAND_CATEGORIES` to disable command groups such as `auto`, and more. You can find the template and explanation of each setting in your `.env.template` file.

When you first start Auto-GPT, you'll be prompted to provide a name, AI role, and goals. These fields are automated by default, meaning you can issue commands directly.

For example, to research Wladastic, the author of *Unlocking the Power of Auto-GPT and its Plugins*, and write the results into a text file, you could issue the following command:

```
"Research Wladastic, the author of Unlocking the Power of Auto-GPT and
Its Plugins and write the results into a text file."
```

Auto-GPT will then try to generate the `ai_settings.yaml` file; if it fails, you will be asked to supply the name of the instance, five main goals of `ai_settings`, and the role that influences the behavior of the instance.

Make sure to be extremely specific and detailed in your prompts. When using Auto-GPT, I tend to edit the `ai_settings.yaml` file manually and it works very well with longer instructions as well as more than 5 goals (this is just a default thing, as it was developed when only GPT-3.5 was available, which has a much lower token limit)

Feel free to research ChatGPT prompting guides to learn how to make Auto-GPT as efficient as possible. Unclear and "too short" prompts may result in Auto-GPT hallucinating or just doing very wrong stuff such as "research for my homework," which may result in various steps such as asking the user (you) about what exactly you want, and this will all generate costs on your OpenAI account.

Summary

In this comprehensive chapter, we delved into the installation and setup of Auto-GPT across various OSs, including Windows, macOS, and Linux, equipping you with the essential knowledge to get started. We began by outlining the system requirements for each platform and provided detailed instructions for installing Python 3.10, which is crucial for running Auto-GPT. Our guide included different methods to obtain Auto-GPT, such as cloning the repository via Git or downloading it from GitHub as a ZIP file.

Once you had Auto-GPT on your system, we led you through its installation using Docker (recommended), Git, or without either. We also explained the process of configuring your `.env` file with your unique OpenAI API key and settings for GPT models in the `FAST_LLM_MODEL` and `SMART_LLM_MODEL` attributes.

After successfully setting up Auto-GPT, we introduced the fundamental concepts of text generation models such as GPT-3/GPT-4 from OpenAI, discussing neural networks, deep learning models for NLP, and the text generation tasks these models perform.

The chapter further explored additional Auto-GPT features, including plugins that enhance its functionality, headless browsers for automated tasks, workspaces for file management, and API keys for secure access to OpenAI's services.

Finally, we demonstrated running your first AI-generated task with Auto-GPT, highlighting its ease of use and power as a tool. We concluded the chapter by preparing you for our next sections, which will dive deeper into advanced Auto-GPT features, such as customization for specific needs and working with various plugins to broaden its capabilities. By mastering these aspects and effectively harnessing the power of AI-generated text, you'll be well equipped for a range of tasks, from automating content creation to generating engaging narratives based on prompts. Stay tuned as we continue exploring the full potential of Auto-GPT in our upcoming chapters.

Building upon the foundational knowledge acquired about installing and configuring Auto-GPT, as well as understanding text generation models, the next chapter we will explore is titled *Mastering Prompt Generation and Understanding How Auto-GPT Generates Prompts*. This chapter promises to deepen your understanding of prompt generation, a crucial skill for maximizing the potential of Auto-GPT. It will demystify the mechanics behind Auto-GPT's prompt generation and offer guidance on crafting effective prompts to enhance your interactions with this advanced language model.

Mastering Prompt Generation and Understanding How Auto-GPT Generates Prompts

You are really into the book! Congratulations on reaching this chapter!

In the previous chapters, we explored the basics of Auto-GPT and its installation. Now, we are going to delve deeper into one of the most crucial aspects of working with this powerful language model – prompt generation. In this chapter, we will demystify the process of how Auto-GPT generates prompts, understand why they are so important, and learn how to craft effective prompts to get the most out of Auto-GPT. Let's get started!

In this chapter, we will explore these topics:

- What are prompts, and why are they important?
- Tips to craft effective prompts
- An overview of how Auto-GPT generates prompts
- Examples of what works and what confuses GPT

What are prompts, and why are they important?

We usually know how to talk to ChatGPT, which is chatting with it directly. ChatGPT responds directly with its answer to whatever we ask it. But how does this relate to prompts?

The text we send is called a prompt; it can be a question, a statement, a task, or just whatever we want to tell the **large language model (LLM)** . However, this text is not fed into the LLM directly.

The application generally provides the context of the conversation, such as constraints (for example, *You are a helpful assistant. Never argue with the user, and answer requests only if it is ethical and helpful to the user*).

Prompts are the initial inputs that you provide to a language model to generate a response. They can take several forms, such as a question, a statement, or a task.

For example, if you were to ask the model, *What is the weather like today?*, this would be a question prompt.

A statement prompt could be something like *Tell me about the history of Rome,*" while a task prompt might be *Write a short story about a spaceship*. Each type of prompt serves a different purpose and can elicit different types of responses from the model. Understanding how to use these different types of prompts effectively is key to getting the most out of your interaction with Auto-GPT.

Here was a prompt that was sent to Auto-GPT, which someone removed:

please add the text that was here before which was containing the prompt—

With that prompt, we provided the following constraints:

- "Never argue with the user":

  - **Interpretation**: The user is always the source of truth.

  - **Effect**: This constraint emphasizes a user-centric approach where the AI refrains from challenging the user's statements or perspectives. It ensures that the AI's responses are in agreement or neutral to the user's input, but this could potentially limit the depth of interactive dialogue. It may also cause the AI to "role play" and act as if the situation was only a story, trying to respond in a manner that fits the context or even the aforementioned sentence. This may cause Auto-GPT to make wrong assumptions or become blind to untrue information, such as steps to setting up something that are made up and only sound kind of right.

- "You are a helpful assistant":

  - **Role definition**: Clearly establishes the AI's role as an assistant.

  - **Tone and interaction**: Sets a predefined tone and direction for conversations. The AI is programmed to strictly adhere to an assistant's role. However, this might not always align with user expectations. Users may anticipate a more informal, friendly tone and interactive engagement, such as receiving proactive questions such as *"How can I assist you today?"* Instead, the AI's involvement might be limited to reactive responses without creative input.

- "Answer questions only if it is ethical and helpful to the user":

  - **The scope of capabilities**: Imposes significant limitations on the AI's functionality. The AI is programmed to prioritize ethical considerations and user utility in every interaction.

- **An impact on decision-making assistance**: This could lead to an overly cautious approach where the AI refrains from performing tasks that might relieve users of decision-making responsibilities. For example, if faced with complex or morally ambiguous tasks, the AI might opt for minimal engagement rather than comprehensive assistance.

- **An example in practice**: When asked to assist in creative tasks, such as writing a chapter about the varying colors of roses, the AI might limit its response to outlining potential approaches (e.g., suggesting bullet points for the chapter), rather than actively engaging in the creative process itself.

Prompts define the task as well as the context of what the LLM is supposed to answer.

As we discussed in *Chapter 1*, Auto-GPT sends a fairly huge prompt, where it defines the context, commands, and constraints, as well as a `"user"` message that says the following:

```
role": "user",
"content": "Determine which next command to use, and respond using the
format specified above:" :" :"
```

This way, GPT actually plays a role in a hypothetical story that contains the current context of information and the possible commands that only Auto-GPT can execute, responding to the query prompt of the user with what command it thinks it should use next.

This is even more interesting if you consider that when instructing, for example, ChatGPT to do something, it would generally respond with the so-annoying phrase, *"As an AI language model..."*

Phrasing

When we look at the prompts that Auto-GPT uses, we could easily think about rephrasing them, making them shorter to save tokens, or adding more context.

However, each of those operations has a downside.

LLMs are not human; they aren't really an intelligence that understands the input that it is given and what it outputs. They are only trained to generate text that is the most probable, given a certain input.

This means that even though a sentence could have the exact same meaning as another, both sentences may be interpreted completely differently.

Due to the nature of machine learning, the texts that were used to train an LLM pretty much also define how it stores its knowledge.

If the sentence *"I have just sold my car because I just wanted to buy an ice cream"* never appeared in the training data, an LLM would have a harder time understanding the sentence, given that the tokens are not related. "Buy" would be related to "car" and "ice cream," but "car" and "ice cream" would be considerably unrelated.

As GPT-4 is not open source, we have to rely on Llama to understand how models work.

Llama has a few parameters, such as length penalties and uniqueness (i.e., how often the same words appear). These limit the vector length of each embedding, for example.

Embeddings

Embeddings are a crucial part of how Auto-GPT generates prompts. They are essentially a way of representing words and phrases in a numerical form that a model can understand. Each word or phrase is represented as a point in a multidimensional space, where the distance and direction between points can represent the relationship between words or phrases.

For instance, in this multidimensional space, the words "king" and "queen" might be close together, indicating a strong relationship, while "king" and "ice cream" would be further apart, indicating a weaker relationship. This is how a model understands context and can generate relevant responses.

The process of creating these embeddings involves breaking down the input text into tokens, which can be words, parts of words, or even single characters, depending on the language. These tokens are then mapped to vectors in the multidimensional space.

The model then uses these vectors to generate a response. It does this by calculating the probability of each possible next token, based on the current context. The token with the highest probability is selected, and the process is repeated until a complete response is generated.

This is a simplified explanation of the process, and the actual implementation involves a lot more complexity, including the use of attention mechanisms to determine which parts of the input are most relevant, and the use of transformer models to handle long sequences of tokens.

However, the key takeaway is that a model generates prompts by understanding the context of the input and calculating the most probable next token. This is why the phrasing of prompts is so important, as it can greatly influence the model's understanding of the context and, therefore, the generated response.

In the next section, we will look at some tips to craft effective prompts and provide some examples of what works and what confuses Auto-GPT.

Tips to craft effective prompts

The art of crafting effective prompts is a skill that requires a nuanced understanding of a language model's capabilities and limitations. The following guidelines will help you to create prompts that are more likely to yield the desired responses from Auto-GPT:

- **Precision is key**: The specificity of your prompt can significantly influence the relevance of the response. For instance, a vague prompt such as *"Tell me about dogs"* might yield a generic response about dogs. However, a more specific prompt such as *"What are the different breeds of dogs and their characteristics?"* is likely to generate a more detailed and informative response.

- **Clarity and simplicity**: It's crucial to remember that while Auto-GPT is a sophisticated language model, it's not a human. Therefore, it's best to avoid jargon or complex language that could potentially confuse the model. Instead, opt for clear, simple language that the model can easily interpret.

- **Contextual Clues**: One of the key tips to craft effective prompts is to provide sufficient contextual clues. This means giving a model enough information to understand the broader context of the conversation or task. For example, if you were asking the model to write a story set in medieval times, you might provide some context about the setting, characters, and plot before giving the actual prompt. This could be done through a more detailed prompt or by utilizing the conversation history feature of Auto-GPT. By providing sufficient context, you can help the model generate a more relevant and coherent response.

- **Experimentation**: Don't hesitate to experiment with different phrasing and approaches. A slight change in the phrasing of your prompt can sometimes lead to a significantly different response.

Next, we will look at a few examples of effective and ineffective prompts.

Examples of effective and ineffective prompts

To better understand these principles, let's examine some examples of such prompts:

- **Example 1**: "Tell me a joke." This prompt is simple and clear, and Auto-GPT is likely to respond with a joke, demonstrating its ability to generate creative content.

- **Example 2**: "What is the meaning of life?" This prompt is philosophical and broad, which might lead to a vague or generic response, as the model might struggle to provide a concise and meaningful answer.

- **Example 3**: "As a language model, explain the concept of machine learning." This prompt is clear, and specific, and provides context, which will likely result in a detailed explanation of the concept.

- **Example 4**: "Translate the following text into French: 'Hello, how are you?'" This prompt is clear, specific, and task-oriented, which should lead to a correct translation.

In conclusion, understanding the intricacies of how Auto-GPT generates prompts and mastering the art of crafting effective prompts can significantly enhance your interaction with the model. Remember, the key is to be specific, use clear and simple language, provide ample context, and embrace experimentation. With these guidelines in mind, you're well on your way to becoming a proficient user of Auto-GPT.

An overview of how Auto-GPT generates prompts

Here, we will understand the prompt generation process in Auto-GPT.

Auto-GPT's prompt generation process is a sophisticated mechanism that involves a deep understanding of the input context and the calculation of the most probable next token. This process is not just about generating responses but also about setting the stage for the conversation, defining the roles, and establishing the rules of engagement.

Let's delve deeper into this process:

- **Tokenization**: The initial step involves breaking down the input text into tokens, which could be words, parts of words, or even individual characters, depending on the language.

- **Embedding**: Each token is then mapped to a vector in a multidimensional space, creating an "embedding." The position of each vector in this space signifies the meaning of the corresponding token in relation to all other tokens.

- **Contextual understanding**: Auto-GPT uses these embeddings to comprehend the context of the input. It calculates the distance and direction between the vectors, representing the relationships between the tokens.

- **Response generation**: The model then generates a response by calculating the probability of each possible next token, based on the current context. The token with the highest probability is selected, and the process is repeated until a complete response is generated.

- **An attention mechanism**: An attention mechanism is employed to determine which parts of the input are most relevant to the current context. This allows the model to focus on the most important parts of the input when generating a response.

- **Transformer models**: To handle long sequences of tokens, the model uses transformer models. These models can process the tokens in parallel, making them much more efficient than traditional sequential models.

In *Chapter 1*, we discussed the default prompt that Auto-GPT uses, which includes constraints, context, goals, and commands. This default prompt sets the stage for a conversation, defines the roles, and establishes the rules of engagement. For instance, constraints such as *"Never argue with the user"* and *"You are a helpful assistant"* set the tone and direction of the conversation. The context and goals provide a clear understanding of the task at hand, and the commands guide the model's responses.

This is a high-level overview of the process, and the actual implementation involves a lot more complexity. However, the key takeaway is that Auto-GPT generates prompts by understanding the context of the input and calculating the most probable next token. This is why the phrasing of the prompts is so important, as it can greatly influence the model's understanding of the context and, therefore, the generated response.

Examples of what works, and what confuses GPT

Here are some examples of what GPT understands and what it might miss out on:

- **Example 1 – an effective prompt**: Here are the AI settings for this one:

  - **Role**: An AI-powered author and researcher specializing in creating comprehensive, well-structured, and engaging content on Auto-GPT and its plugins, while maintaining an open line of communication with the user for feedback and guidance

  - **Goals**: Conduct a thorough analysis of the current state of the book and identify areas for improvement

  - **Prompt**: AuthorGPT, I have placed a text file in your working directory; can you analyze the current state of the book and suggest areas for improvement?

  This prompt aligns perfectly with the role and goal defined in the AI settings. The model is likely to respond with a detailed analysis of the book and suggestions for improvement.

- **Example 2 – ineffective prompts**: Understanding hallucinations in GPT models – a compact exploration.

  Hallucinations in GPT models refer to occasions where a model generates text that seems logical but is not based on reality. This usually occurs when the model encounters vague or incomplete prompts. Interestingly, generative AI such as GPT can begin to "hallucinate." We'll delve into the mechanics of this phenomenon shortly, but first, let's define it more clearly. Here are the AI settings:

  - **Role**: An AI-powered author and researcher specializing in creating comprehensive, well-structured, and engaging content on Auto-GPT and its plugins, while maintaining an open line of communication with the user for feedback and guidance.

  - **Goals**: Conduct a thorough analysis of the current state of the book and identify areas for improvement

  - **Prompt**: AuthorGPT, can you analyze the current state of the book and suggest areas for improvement?

With this prompt, you would expect Auto-GPT to ask you for the context, but I found that GPT tries to improvise instead and starts to hallucinate.

Hallucination with GPT means it starts to act as if it is doing something factual whereas it isn't. Let's understand this more in depth ahead.

What does hallucination mean in GPT?

Hallucination in GPT models manifests when a model acts as though it's performing a task. This ranges from creating code files for projects that only contain placeholders to fabricating facts that sound contextually appropriate. For example, in certain situations, it might discuss a topic closely related to the context's keywords.

Hallucination in language models such as GPT happens when the model produces text that appears plausible but is unanchored in reality. This is often the result of the model working with ambiguous or insufficient information.

Take, for instance, if you request the model to write a story about a non-existent character. It might "hallucinate" details about the character's life, appearance, or traits. While this can yield creative and unforeseen results, it can also lead to text that is illogical or unrelated to the original prompt.

A case in point is when I asked Auto-GPT to develop a `three.js`-based RPG browser game. It began researching how to gather weather data from a non-existent API. This was a result of processing excessive context with GPT-3.5-turbo, which I had used before transitioning to GPT-4.

Auto-GPT's memory sometimes harbors incorrect facts or memories. To economize tokens, its memory is condensed by *Chat Completion* prompts with GPT. This can lead to misunderstandings, especially when summarizing texts that have already been summarized and merging these summaries.

Confusion can also stem from tokens closely related to the current context. For example, if writing a weather API tool and a web-based game are both related to "JavaScript," the model might perceive the weather API as a relevant topic. Such confusion is less frequent with GPT-4, thanks to its advanced parameters and enhanced precision.

Confusing a prompt and its impact on AI performance

Here are some AI settings where the prompt can be confusing, impacting the performance of the AI it works with:

- **Role**: An AI-powered author and researcher specializing in creating comprehensive, well-structured, and engaging content on Auto-GPT and its plugins, while maintaining an open line of communication with the user for feedback and guidance

- **Goals**: Conduct a thorough analysis of the current state of the book and identify areas for improvement

- **Prompt**: AuthorGPT, can you tell me a joke?

  This seemingly simple prompt can lead to confusion for the model, primarily because it diverges from its established role and goal. The request for a joke seems out of place in the context of analyzing a book and suggesting enhancements.

Initially, the model might comply and tell a joke. However, this deviation can have longer-term repercussions. As the model integrates this interaction into its memory summary, it may become increasingly perplexed. This is because Auto-GPT instructs GPT to retain as much information as possible to prevent topic shifts, hallucinations, or previous steps being forgotten.

A specific issue arises when Auto-GPT focuses on a particular task but encounters an unrelated prompt. For example, if it receives a command irrelevant to the ongoing context, the model might lose track of its previous actions. It could end up in a loop of searching for information related to the new input, attempting to reconcile it with the earlier task. As a result, Auto-GPT might start intertwining the unrelated joke with its Google Search results, leading to a mix-up of topics.

This scenario highlights the importance of aligning prompts with the AI's defined role and objectives to maintain effectiveness and avoid confusion.

An effective prompt and its impact on AI performance

Here are the AI settings for an effective prompt and its advantages on AI performance:

- **Role**: An AI-powered author and researcher specializing in creating comprehensive, well-structured, and engaging content on Auto-GPT and its plugins, while maintaining an open line of communication with the user for feedback and guidance

- **Goal**: Develop a comprehensive plan to create task lists that will help you structure research, a detailed outline per chapter, and individual parts

- **Prompt**: AuthorGPT, can you help me develop a comprehensive plan to create task lists to structure my research and outline for each chapter?

This prompt is clear, specific, and aligns perfectly with the role and goal defined in the AI settings. The model is likely to provide a detailed plan for creating task lists and structuring research.

Confusing a prompt and its impact on AI performance

Here is an example of a confusing prompt and its impact on AI performance:

- **Role**: An AI-powered author and researcher specializing in creating comprehensive, well-structured, and engaging content on Auto-GPT and its plugins, while maintaining an open line of communication with the user for feedback and guidance

- **Goal**: Develop a comprehensive plan to create task lists that will help you structure research, a detailed outline per chapter, and individual parts.

- **Prompt**: AuthorGPT, what is the weather like today?

Understanding the disparity

This prompt stands in stark contrast to AI's designated role and goal. The model, configured to focus on task list creation and research structuring, faces a dilemma with a prompt that is unrelated to these tasks.

This is likely to confuse the model because it doesn't align with the defined role and goal. The model might struggle to provide a relevant response because the prompt doesn't involve creating a plan or structuring research.

Here are some potential AI behavior scenarios:

- While Auto-GPT may respond correctly by researching the weather or just shrugging off that question by explaining that it is not the main focus, it could get confused and drop the previous task, either partially or completely. This could result in a very inaccurate future behavior, or even the issue escalating to GPT not responding correctly to the Auto-GPT module that communicates with it, causing a fatal error.

- The model might also try to relate every decision to checking the weather now, or it will keep checking the weather in later steps if Auto-GPT does not manage to compress its memory correctly (for example, when it tries to summarize the memory to reduce data amounts, it may put more emphasis on weather data now).

In conclusion, crafting effective prompts for Auto-GPT involves aligning your prompts with the defined role and goal in the AI settings. Clear, specific prompts that align with these parameters are more likely to yield relevant responses, while prompts that don't align can confuse the model.

Summary

In this chapter, we delved into the intricacies of prompt generation and how Auto-GPT generates prompts. We started by defining prompts and their importance in shaping the responses of the language model. We learned that prompts can be questions, statements, tasks, or any text that we want to communicate to a language model.

We also discussed the role of constraints in providing context to a conversation and guiding a model's responses. We examined how specific constraints can influence the tone, direction, and ethical boundaries of the conversation.

We then explored the technical aspects of prompt generation, including tokenization, embedding, context understanding, response generation, attention mechanisms, and transformer models. We learned that a model generates prompts by understanding the context of the input and calculating the most probable next token.

Then, we provided tips to craft effective prompts, emphasizing the importance of specificity, clarity, context, and experimentation. We also looked at examples of effective and confusing prompts, demonstrating how alignment with the defined role and goal in the AI settings can influence a model's responses.

Finally, we examined examples of prompts based on specific AI settings, demonstrating how effective prompts align with the defined role and goal, while confusing prompts do not.

In conclusion, mastering prompt generation and understanding how Auto-GPT generates prompts can significantly enhance your interaction with a model. The key is to craft clear, specific prompts that align with the model's role and goal, provide ample context, and not be afraid to experiment.

In the next chapter, we will use the skills we acquired with plugins and learn how to customize prompts.

4

Short Introduction to Plugins

Welcome to *Chapter 4*! In this chapter, we will provide a brief introduction to **plugins** in **Auto-GPT**.

As more and more people came up with new ideas on what Auto-GPT could do, we realized that it would be impossible to implement all of them into the core project of Auto-GPT, so we decided to create a plugin system that allows users to extend the functionality of Auto-GPT by adding their own plugins.

In this chapter, we will go through these topics:

- Going through an overview of plugins in Auto-GPT
- Knowing the types of plugins and their use cases
- Learning how to use plugins
- Understanding how plugins are built
- Using my Telegram plugin as a hands-on example

Going through an overview of plugins in Auto-GPT

Plugins in Auto-GPT serve as modular extensions that enable additional functionality and customization of your own Auto-GPT instance. They provide a way to integrate external tools, services, and models seamlessly into the Auto-GPT framework. By leveraging plugins, you can tailor Auto-GPT to suit specific tasks, domains, or applications, such as having your own customer support chat, having your own researching AI that gives you suggestions or helps you schedule your calendar, and so much more!

Auto-GPT has interfaces that allow you to integrate it with almost anything that has text output or non-visual interface (with a little bit of coding, you may even make a VS Code plugin that allows Auto-GPT to navigate through projects and lines of code).

Auto-GPT has an official plugin repository that contains a wide range of plugins. These plugins were developed as standalone plugins at first, but anyone who wants to add their plugin to the official list can do so by submitting a pull request. The official plugins are maintained by the Auto-GPT team and are thoroughly tested to ensure compatibility with the latest version of Auto-GPT. Although the original creator is responsible for maintaining the plugin, the Auto-GPT team will provide support and assistance as needed but also remove plugins that are no longer maintained and do not work anymore.

To help everyone get started with creating a plugin, the Auto-GPT team has created a plugin template that can be used as a starting point for creating your own plugin. The template contains all the necessary files and folders required for a plugin, including a `README` file with instructions on how to use it. The template is available on GitHub and can be downloaded from the repository at `https://github.com/Significant-Gravitas/Auto-GPT-Plugin-Template`.

Knowing the types of plugins and their use cases

Plugins can be created for any purpose, given that the plugin template allows it. There are various types of plugins available in `Auto-GPT-Plugins`, each catering to different use cases.

Here are a few examples of official plugins:

- **Astro Info**: This provides Auto-GPT with information about astronauts
- **API Tools**: This allows Auto-GPT to make API calls of various kinds
- **Baidu Search**: This integrates Baidu search engines into Auto-GPT
- **Bing Search**: This integrates Bing search engines into Auto-GPT
- **Bluesky**: This enables Auto-GPT to retrieve posts from Bluesky and create new posts
- **Email**: This automates email drafting and intelligent replies using AI
- **News Search**: This integrates news article searches, using the NewsAPI aggregator, into Auto-GPT
- **Planner**: This provides a simple task planner module for Auto-GPT
- **Random Values**: This enables Auto-GPT to generate various random numbers and strings
- **SceneX**: This explores image storytelling beyond pixels with Auto-GPT
- **Telegram**: This provides a smoothly working Telegram bot that gives you all the messages you would normally get through the Terminal
- **Twitter**: This retrieves Twitter posts and other related content by accessing the Twitter platform via the v1.1 API using Tweepy
- **Wikipedia Search**: This allows Auto-GPT to use Wikipedia directly
- **WolframAlpha Search**: This allows Auto-GPT to use WolframAlpha directly

The community is also coming up with new plugins all the time. These are then found in the official Auto-GPT Discord server in the **#plugins** channel:

- **Language model (LM) plugins**: LM plugins allow you to incorporate specialized LMs into Auto-GPT. These plugins enable fine-tuned models that are designed for specific tasks or domains, such as code generation, translation, summarization, sentiment analysis, and more.

- **Data source plugins**: Data source plugins enable Auto-GPT to access external data sources and retrieve information on demand. These plugins can connect Auto-GPT to databases, APIs, web scraping tools, or other data repositories. With data source plugins, you can enrich Auto-GPT's knowledge and enable it to provide up-to-date and relevant information to users.

- **Chatbot plugins**: Chatbot plugins facilitate interactive and dynamic conversations with Auto-GPT. These plugins incorporate dialogue management techniques, allowing Auto-GPT to maintain context, remember previous interactions, and generate coherent responses. Chatbot plugins are useful for building chat assistants, customer support bots, virtual companions, and more.

Learning how to use plugins

Using plugins in Auto-GPT can be a bit tricky at first, but once you get the hang of it, it's pretty easy.

Auto-GPT has a `plugins` folder in its root directory, where all the plugins are stored.

The method of how the plugins are installed has changed a bit over time – you can either clone the desired plugin repository into the `plugins` folder and zip it or just leave it there and Auto-GPT will find it.

By the time you read this book, the plugin system might be updated to use a plugin manager, which will make it easier to install and manage plugins.

After placing the plugin, you have to install the dependencies that are required for the plugin to work. This can be done by running the following command while starting Auto-GPT:

```
python -m autogpt --install-plugin-deps
```

Sometimes, this function does not work. If the plugin is not installed, navigate to the `plugins` folder and into the folder of your plugin run the following command:

```
pip install -r requirements.txt
```

Auto-GPT should automatically install the dependencies for the plugin, and it should now tell you that it found the plugin. It should also tell you that the plugin is not configured yet and that you should configure it.

Some of the plugins may still indicate that you have to make changes to the .env file, but currently, the .env file is not used anymore, so you have to configure the plugin config file, or if by the release of this book, we have finished the plugin manager, you can configure the plugin via the plugin manager.

To figure out what names to use, just start Auto-GPT normally and it will list the plugins it found. If it wasn't configured automatically, it will tell you that it is not.

If by the time you read this book, the architecture of the plugin system has changed, you can find the information on how to configure the plugin in the README.md file of the Auto-GPT-Plugins repository.

Understanding how plugins are built

Plugins in Auto-GPT are built using a modular and extensible architecture. The exact process of building a plugin may vary depending on the type and complexity of the plugin.

Structure of a plugin

The plugin should be in its own folder and contain a __init__.py, which contains the AutoGPTPluginsTemplate class reference. Each of the class methods contains a method that determines whether the following method is active, for example:

post_prompt is only active if can_ post_prompt returns True.

As we are limited to the plugin template, we can only use the methods that the template provides. Each method has a can_handle method that returns a boolean value, which is used to determine whether the plugin can handle the current prompt or method. The plugin methods are scattered all over the Auto-GPT code and allow plugins to add functions as commands that Auto-GPT can cognitively call to give Auto-GPT agents new abilities.

Here are some of these interface methods:

- post_prompt: This adds access to the prompt generator. This allows the plugins to edit the prompts or add new functions as commands.
- on_response: This forwards the content of a chat completion response to the plugin and returns the edited content to Auto-GPT.
- on_planning: This allows the plugin to edit the message sequence before it is sent to the Auto-GPT agent, for example, to summarize the history of the messages or to add a new message to the sequence.
- post_planning: This allows the editing of the response JSON of the agent's thought planning. For example, this could be used to add another step of thought such as reevaluating the decision of the agent and which command it chose to execute.

- `pre_command`: After the user approves the command the agent chooses, the plugin can edit the command before it is executed.

- `post_command`: After the command has been executed and right before the results of that command are returned to the user, the plugin can edit the result and also has access to the command name that was executed.

- `handle_chat_completion`: This is used to add a custom chat completion function to Auto-GPT agents. If this is enabled, OpenAI's GPT will not be used for chat completion mostly but could still be used if at certain places only GPT can be used or just not implemented cleanly.

- `handle_text_embedding`: This adds the ability to add text embedding to Auto-GPT agents other than the memory module.

- `user_input`: This is used to forward the user input query to the plugin instead of the console or terminal.

- `report`: This is used to forward logs to the plugin that would usually only be printed to the console or terminal.

You are also free to copy and paste parts of other plugins that you find useful and use them in your own plugin, as long as you give credit to the original author of the code.

The Planner plugin is a good example of how to use the `PromptGenerator` class to add new commands to Auto-GPT agents.

If you want to create a plugin that enables communication, you may also check what already exists in the Auto-GPT Discord.

Multiple projects exist now that also enable multiple ways to communicate with Auto-GPT and maybe you can even come up with the ultimate communication plugin that enables Auto-GPT to communicate with their humans in any way possible.

How to build plugins

As you embark on the journey of building plugins for Auto-GPT, it's crucial to approach the process with a well-rounded strategy. Here, we outline a step-by-step guide to help you plan, develop, and share your plugin effectively within the community.

It is a good idea to plan out what you want to do with your plugin and what you want to achieve with it. A basic procedure would be as follows:

1. **Define functionality**: Start by defining the functionality or features the plugin will provide. Identify the specific task, domain, or integration point that the plugin aims to address.

2. **Implement plugin logic**: Develop the necessary code to implement the desired functionality. This may involve writing custom classes, functions, or methods that interact with Auto-GPT or external services.

3. **Handle integration**: Consider how the plugin will integrate with Auto-GPT. This may involve hooking into specific events or methods within the Auto-GPT framework to intercept prompts, modify responses, or access data sources.

4. **Test and refine**: Thoroughly test the plugin to ensure its functionality and compatibility with Auto-GPT. Iterate and refine the plugin based on feedback and testing results. If you can, write unit tests for your plugin to ensure that it works as intended.

    Although the team is very forthcoming, if you want your plugin to become an official "first party" plugin, you should write unit tests for it and document it properly. Otherwise, if people don't understand your plugin and it is not documented properly, it will not stay in the official plugin repository for long as it requires more time to read into the code to check how it works.

5. **Publish and share**: Once the plugin is ready, you can publish and share it with the Auto-GPT community. This allows other users to benefit from your plugin and encourages collaboration and innovation within the ecosystem. Most people just create a Discord thread in the **#plugins** channel and share their plugin there, so if someone has a question, you can answer directly without having to handle multiple GitHub issues being created (which can be frustrating, as once you resolve an issue, people close it and the next day another person will ask the same question and most users do not even provide enough information of what they did or they just do not read the README.md file of the plugin or just skip steps and just tell you stuff that is nowhere related to your plugin).

Now that we have learned a simple way of planning a plugin that would be then qualified to join Auto-GPT's plugin list, we can use one of the plugins as an example to build one ourselves.

Using my Telegram plugin as a hands-on example

Here, we will go through the plugin example and see the steps we need to follow:

1. To demonstrate how plugins are made, I decided to include my Auto-GPT plugin for Telegram integration.

2. It involves simply forwarding messages to the user and is also capable of asking the user a question and awaiting a response. Basically, your Telegram chat becomes a remote extension to the console/terminal application and you can leave Auto-GPT running on your machine and operate it remotely with your phone.

3. Fill the interface class in the `__init__.py` file. This file acts as the core of the `AutoGPTTelegram` plugin. It houses the `AutoGPTTelegram` class, which inherits from `AutoGPTPluginTemplate`. To get the template, go to the **Significant Gravitas GitHub** page and either clone the Auto-GPT plugin template (which by this time is outdated) or copy-paste any of the other plugin folders and make your own by just modifying `__init__.py` and return `False` on the methods you do not intend to use.

4. The __init__ method is crucial for setting up the plugin. It initializes a TelegramUtils object that will be used for interactions with the Telegram API:

```
class AutoGPTTelegram(AutoGPTPluginTemplate):
    def __init__(self):
        super().__init__()
        self._name = "AutoGPTTelegram"
        self._version = "1.0.0"
        self._description = "This plugin integrates Auto-GPT
with a Telegram bot."
        self.telegram_chat_id = "YOUR_TELEGRAM_CHAT_ID"
        self.telegram_api_key = "YOUR_TELEGRAM_API_KEY"
        self.telegram_utils = TelegramUtils(
            chat_id=self.telegram_chat_id,
            api_key=self.telegram_api_key
    )
```

Here, self._name, self._version, and self._description are attributes to describe the plugin while self.telegram_chat_id and self.telegram_api_key are placeholders for your Telegram credentials. The TelegramUtils object is created with these credentials.

5. The can_handle_user_input and user_input methods work in tandem to handle user input:

```
def can_handle_user_input(self, user_input: str) -> bool:
    return True

def user_input(self, user_input: str) -> str:
    return self.telegram_utils.ask_user(prompt=user_input)
```

The can_handle_user_input method returns True, indicating that this plugin can handle user input. The user_input method takes the user's input and calls the ask_user method of TelegramUtils to interact with the user via Telegram.

6. The can_handle_report and report methods are designed to manage reporting:

```
def can_handle_report(self) -> bool:
    return True

def report(self, message: str) -> None:
    self.telegram_utils.send_message(message=message)
```

Similar to user input handling, can_handle_report returns True to signify that this plugin can handle reporting. The report method sends a message to the user via Telegram using the send_message method of TelegramUtils.

7. Other methods in this class are disabled by default but can be enabled to extend functionality:

```
def can_handle_on_response(self) -> bool:
return False
```

The `can_handle_on_response` method here is a placeholder that could be enabled to process responses in a certain way.

8. The `telegram_chat.py` file contains the `TelegramUtils` class, which houses utility methods for Telegram interactions. Of course, you could write all you need in the `init` file, but it would be less than readable in the end. This walk-through might even be chunked into more files, but as I try to cover as many readers with different levels of knowledge as possible, we only do two files in total.

    I.   We will first write a `TelegramUtils` class:

```
class TelegramUtils:
    def __init__(self, api_key: str = None,
        chat_id: str = None):
# this is filled in the next step.
```

    II. The `__init__` method in the `TelegramUtils` class initializes the `TelegramUtils` object with the API key and chat ID or guides the user on how to obtain them if they're not provided:

```
def __init__(self, api_key: str = None,
    chat_id: str = None):
    self.api_key = api_key
    self.chat_id = chat_id
    if not api_key or not chat_id:
        # Display instructions to the user on how to get
API key and chat ID
        print("Please set the TELEGRAM_API_KEY and
        TELEGRAM_CHAT_ID environment variables.")
```

Here, if `api_key` or `chat_id` is not provided, instructions are displayed to the user on how to obtain them.

In the actual plugin, I decided to add more information for the user; the `__init__` method of the `TelegramUtils` class is more extensive and the code further handles the scenario where `api_key` or `chat_id` is not provided:

```
if not api_key:
    print("No api key provided. Please set the
        TELEGRAM_API_KEY environment   variable.")
    print("You can get your api key by talking to @
        BotFather on Telegram.")
```

```
            print( "For more information:
https://core.telegram.org/bots/tutorial#6-  b  otfather"  )
            return


        if not chat_id:
            print( "Please set the TELEGRAM_CHAT_ID
                environment variable.")
            user_input = input( "Would you like to send a
test message to your bot to get the   id? (y/n): ")
            if user_input == "y":
                try:
                    print("Please send a message to your
telegram bot now.")
                    update = self.poll_anyMessage()
                    print("Message received!
                        Getting chat id...")
                    chat_id = update.message.chat.id
                    print("Your chat id is: " +
                        str(chat_id))
                    print("And the message is: " +
                        update.message.text)
                    confirmation =
                        random.randint(1000, 9999)
                    print("Sending confirmation message: "
                        + str(confirmation))
                    text = f"Chat id is: {chat_id} and the
                      confirmation code is {confirmation}"
                    self.chat_id = chat_id
                    self.send_message(text)
# Send confirmation message
                    print( "Please set the TELEGRAM_CHAT_ID
                        environment variable to this.")
                except TimedOut:
                    print( "Error while sending test
                    message. Please check your Telegram
                        bot.")
                    return
        self.chat_id = chat_id
```

In the preceding block, the method first checks whether api_key is provided. If not, it instructs the user to set the TELEGRAM_API_KEY environment variable and provides guidance on where to obtain the API key. Similarly, for chat_id, it instructs the user to set the TELEGRAM_CHAT_ID environment variable and offers to send a test message to the bot to obtain the chat ID if the user agrees.

III.   The `ask_user` method is designed to prompt the user for input through Telegram. It calls its asynchronous counterpart, `ask_user_async`, to handle the user input asynchronously:

```
def ask_user(self, prompt):
    try:
        return asyncio.run(
            self.ask_user_async(prompt=prompt))
    except TimedOut:
        print("Telegram timeout error, trying again...")
        return self.ask_user(prompt=prompt)
```

Here, the `ask_user` method calls `ask_user_async` within a try block to handle any `TimedOut` exceptions that might occur during the process.

IV.   The `user_input` method handles user input within the plugin, using the `telegram_utils.ask_user` method to gather input from the user via Telegram:

```
def user_input(self, user_input: str) -> str:
    user_input = remove_color_codes(user_input)
    try:
        return self.telegram_utils.ask_user(
            prompt=user_input)
    except Exception as e:
        print(e)
        print("Error sending message to telegram")
        return "s"  # s means that auto-gpt should rethink
its last step, indicating an error with the call
```

The `user_input` method first sanitizes the input to remove color codes, then calls the `ask_user` method of `TelegramUtils` to interact with the user on Telegram.

V.   Write the `report` method in the `AutoGPTTelegram` class to send messages. This method is used for sending status reports or any other messages from Auto-GPT to the user via Telegram.

```
def report(self, message: str) -> None:
    message = remove_color_codes(message)
    try:
        self.telegram_utils.send_message(message=message)
    except Exception as e:
        print(e)
        print("Error sending message to telegram")
```

In this method, any color codes in the message are removed first, and then the `send_message` method of `TelegramUtils` is called to send the message to the user on Telegram.

9. Moving on to the `telegram_chat.py` file, it contains the `TelegramUtils` class, which encapsulates the following utility methods for Telegram interactions:

-   The `__init__` method in the `TelegramUtils` class, which has already been explained.

-   Implementing the `get_bot` method, which is responsible for obtaining a Telegram bot instance using the bot token:

```python
async def get_bot(self):
    bot_token = self.api_key
    bot = Bot(token=bot_token)
    commands = await bot.get_my_commands()
    if len(commands) == 0:
        await self.set_commands(bot)
    commands = await bot.get_my_commands()
    return bot
```

In this method, a new bot instance is created using the `Bot` class from the Telegram package. The `get_bot` method checks whether the bot has any commands set already, and if not, it calls `set_commands` to set the commands for the bot.

-   Implementing `poll_anyMessage` and `poll_anyMessage_async` methods, which are designed to poll for any message sent to the bot:

```python
def poll_anyMessage(self):
    loop = asyncio.new_event_loop()
    asyncio.set_event_loop(loop)
    return loop.run_until_complete(self.poll_anyMessage_async())

async def poll_anyMessage_async(self):
    bot = Bot(token=self.api_key)
    last_update = await bot.get_updates(timeout=30)
    if len(last_update) > 0:
        last_update_id = last_update[-1].update_id
    else:
        last_update_id = -1
    while True:
        try:
            print("Waiting for first message...")
            updates = await bot.get_updates(
                offset=last_update_id + 1, timeout=30)
            for update in updates:
                if update.message:
                    return update
        except Exception as e:
```

```
            print(f"Error while polling updates: {e}")
         await asyncio.sleep(1)
```

Here, `poll_anyMessage` sets up a new `asyncio` event loop and calls `poll_anyMessage_async` to poll for messages asynchronously.

- Implementing the `send_message` and `_send_message` methods, which are utilized for sending messages to the Telegram chat:

```
def send_message(self, message):
    try:
        loop = asyncio.get_running_loop()
    except RuntimeError as e:
        loop = None


    try:
        if loop and loop.is_running():
            print(
                "Sending message async, if this fails its due to
rununtil complete task"
            )
            loop.create_task(
                self._send_message(message=message))
        else:
            eventloop = asyncio.get_event_loop
            if hasattr(eventloop, "run_until_complete")
            and eventloop.is_running():
                print("Event loop is running")
                eventloop.run_until_complete(
                    self._send_message(message=message))
            else:
                asyncio.run(self._send_message(message=message))
    except RuntimeError as e:
        print(traceback.format_exc())
        print("Error while sending message")
        print(e)

async def _send_message(self, message):
    print("Sending message to Telegram.. ")
    recipient_chat_id = self.chat_id
    bot = await self.get_bot()

    # properly handle messages with more than 2000 characters by
chunking them
    if len(message) > 2000:
```

```
        message_chunks = [
            message[i : i + 2000] for i in range(0,
                len(message), 2000)
        ]
        message_chunks = [
            message[i : i + 2000] for i in range(0,
                len(message), 2000)
        ]
        for message_chunk in message_chunks:
            await bot.send_message(
                chat_id=recipient_chat_id,
                text=message_chunk)
    else:
        await bot.send_message(
            chat_id=recipient_chat_id, text=message)
```

In send_message, it first tries to get the current running asyncio event loop. If there isn't one running, it sets loop to None. _send_message is the asynchronous counterpart that actually sends the message to Telegram.

- Implementing ask_user, ask_user_async, and _poll_updates methods, which manage the interaction of asking the user a question and waiting for their response on Telegram:

```
    async def ask_user_async(self, prompt):
        global response_queue

        response_queue = ""
        # await delete_old_messages()

        print("Asking user: " + question)
        await self._send_message(message=question)
        await self._send_message(message=question)

        print("Waiting for response on Telegram chat...")
        await self._poll_updates()

        response_text = response_queue

        print("Response received from Telegram: " +
            response_text)
        return response_text

async def _poll_updates(self):
    global response_queue
    bot = await self.get_bot()
```

```python
        print("getting updates...")
        try:
            last_update = await bot.get_updates(timeout=1)
            if len(last_update) > 0:
                last_update_id = last_update[-1].update_id
            else:
                last_update_id = -1

            print("last update id: " + str(last_update_id))
            while True:
                try:
                    print("Polling updates...")
                    updates = await bot.get_updates(
                        offset=last_update_id + 1, timeout=30)
                    for update in updates:
                        if update.message and update.message.text:
                            if self.is_authorized_user(update):
                                response_queue = update.message.text
                                return
                        last_update_id = max(
                            last_update_id, update.update_id)
                except Exception as e:
                    print(f"Error while polling updates: {e}")
                await asyncio.sleep(1)
        except RuntimeError:
            print("Error while polling updates")

def ask_user(self, prompt):
    print("Asking user: " + prompt)
    try:
        loop = asyncio.get_running_loop()
    except RuntimeError:  # 'RuntimeError: There is no current
event loop...'
        loop = None
    try:
        if loop and loop.is_running():
            return loop.create_task(
                self.ask_user_async(prompt=prompt))
        else:
            return asyncio.run(
                self.ask_user_async(prompt=prompt))
    except TimedOut:
        print("Telegram timeout error, trying again...")
        return self.ask_user(prompt=prompt)
```

In `ask_user_async`, a question is sent to the user on Telegram and `_poll_updates` is called to wait for their response. The `ask_user` method serves as a synchronous wrapper around `ask_user_async`.

Each of these methods plays a critical role in the Telegram interaction, allowing Auto-GPT to communicate with the user via a Telegram bot. The process is well structured, ensuring that the plugin can handle various scenarios that might arise during the interaction.

The methods and code snippets discussed thus far provide a well-rounded framework for integrating Auto-GPT with a Telegram bot. The `telegram_chat.py` file encapsulates Telegram-specific logic, while the `__init__.py` file handles the interaction with Auto-GPT using the `TelegramUtils` class.

Now, let's delve further into some specific segments of the code that may require additional elaboration:

- **Handling long messages**: In the `_send_message` method, there's a segment of code dedicated to handling messages that exceed 2000 characters:

```
if len(message) > 2000:
    message_chunks = [
        message[i : i + 2000] for i in range(0,
            len(message), 2000)
    ]
    for message_chunk in message_chunks:
        await bot.send_message(
            chat_id=recipient_chat_id,
            text=message_chunk)
else:
    await bot.send_message(
        chat_id=recipient_chat_id, text=message)
```

This segment ensures that if a message is longer than 2,000 characters, it's split into chunks of 2,000 characters each, and each chunk is sent as a separate message to Telegram. This is essential for ensuring the integrity of the message being sent, given the maximum message length constraint in Telegram.

- **Re-trying on timeout**: In the `ask_user` method, there's logic to handle a `TimedOut` exception by re-trying the `ask_user` method:

```
except TimedOut:
    print("Telegram timeout error, trying again...")
    return self.ask_user(prompt=prompt)
```

This is a robust way to handle timeouts, ensuring that the plugin retries to ask the user for input until successful.

- **Asynchronous handling**: Various parts of the code utilize asynchronous programming principles to ensure non-blocking operations, such as the following:

```
async def _send_message(self, message):
    # ...
    await bot.send_message(chat_id=recipient_chat_id,
        text=message)
```

Utilizing asynchronous methods such as `await bot.send_message(...)` ensures that the IO-bound operations do not block the execution of the program, leading to a more responsive and performant plugin.

- **Error handling**: Throughout the code, exception handling is employed to catch and handle errors gracefully, ensuring that any issues are logged and dealt with appropriately:

```
except Exception as e:
    print(f"Error while polling updates: {e}")
```

This approach promotes robustness and error resilience in the plugin's operation.

The walk-through of the code has covered the fundamental aspects of how the Telegram plugin for Auto-GPT operates, from initialization to user interaction and error handling. However, there are still some nuanced elements and potential enhancements that could be considered to refine or expand the plugin's functionality. Here are a few additional points and recommendations:

- **Authorization check**: The code snippet provided doesn't include an implementation for the `is_authorized_user` method, which is called in `_poll_updates`. Implementing authorization checks can be crucial to ensure that the bot only responds to messages from authorized users:

```
def is_authorized_user(self, update):
    # authorization check based on user ID or username
    return update.effective_user.id == int(self.chat_id)
```

- **Command handling**: While the `get_bot` method mentions setting commands for the bot, the `set_commands` method is not shown in the provided snippets. It's advisable to implement command handling to provide users with a guide on how to interact with the bot:

```
async def set_commands(self, bot):
    await bot.set_my_commands(
        [
            ("start", "Start Auto-GPT"),
            ("stop", "Stop Auto-GPT"),
            ("help", "Show help"),
            ("yes", "Confirm"),
            ("no", "Deny"),
            ("auto", "Let an Agent decide"),
```

```
        ]
    )
```

Of course, we would also have to modify the `ask_user` method to handle the commands, but this is just a basic example of how to implement command handling:

```python
async def ask_user_async(self, prompt):
    global response_queue
    # only display confirm if the prompt doesn't have the
    string ""Continue (y/n):"" inside
    if "Continue (y/n):" in prompt or "Waiting for your
    response..." in prompt:
        question = (
            (
            prompt
            + " \n Confirm: /yes      Decline: /no \n Or type
    your answer. \n or press /auto to let an Agent decide."
            )
        )
    elif "I want Auto-GPT to:" in prompt:
        question = prompt
    else:
        question = (
            (
            prompt + " \n Type your answer or press /auto to
    let an Agent decide."
            )
        )

    response_queue = ""
    # await delete_old_messages()

    print("Asking user: " + question)
    await self._send_message(message=question)
    await self._send_message(message=question)

    print("Waiting for response on Telegram chat...")
    await self._poll_updates()

    if response_queue == "/start":
        response_queue = await self.ask_user(
            self,
            prompt="I am already here... \n Please use /stop
    to stop me first.",
        )
    if response_queue == "/help":
```

```
            response_queue = await self.ask_user(
                self,
                prompt="You can use /stop to stop me \n and /
start to start me again.",
            )
        if response_queue == "/auto":
            return "s"
        if response_queue == "/stop":
            await self._send_message("Stopping Auto-GPT now!")
            await self._send_message("Stopping Auto-GPT now!")
            exit(0)
        elif response_queue == "/yes":
            response_text = "yes"
            response_queue = "yes"
        elif response_queue == "/no":
            response_text = "no"
            response_queue = "no"
        if response_queue.capitalize() in [
            "Yes",
            "Okay",
            "Ok",
            "Sure",
            "Yeah",
            "Yup",
            "Yep",
        ]:
            response_text = "y"
        elif response_queue.capitalize() in ["No", "Nope",
            "Nah", "N"]:
            response_text = "n"
        else:
            response_text = response_queue
        print("Response received from Telegram: "
            + response_text)
        return response_text
```

- **Logging**: Incorporating a logging framework as opposed to using print statements would provide a more robust and configurable way to log messages and errors. I first tried using the built-in logging of Auto-GPT but importing code from Auto-GPT into the plugin caused some issues over time, so I decided to use the built-in logging module of Python instead:

```
import logging

logging.basicConfig(level=logging.INFO)
```

```
logger = logging.getLogger(__name__)
```

Here is an example of its usage:

```
logger.info("Information message")
logger.error("Error message")
```

- **Environment variable management**: The code retrieves the Telegram API key and chat ID directly. It's a good practice to manage such sensitive information using environment variables, ensuring they are not hardcoded:

```
import os

TELEGRAM_API_KEY = os.getenv("TELEGRAM_API_KEY")
TELEGRAM_CHAT_ID = os.getenv("TELEGRAM_CHAT_ID")
```

- **Code modularity and reusability**: It might be beneficial to further modularize the code, separating concerns and making it easier to maintain and extend. For instance, the Telegram interaction logic could be encapsulated into a separate module or class, making the code more organized and reusable.

- **Unit testing**: Adding unit tests to verify the functionality of the plugin is crucial for ensuring its reliability and ease of maintenance, especially when changes or updates are made to the code base.

- **Documentation**: Ensuring that the code is well documented, including comments explaining the functionality of methods and complex code segments, will make it easier for others to understand, use, and potentially contribute to the plugin.

By considering these additional points and recommendations, developers can enhance the Telegram plugin's functionality, making it more robust, user friendly, and maintainable. Furthermore, readers and developers following the guide will have a more comprehensive understanding of the considerations involved in building and refining a plugin for Auto-GPT.

The discussion thus far has provided a comprehensive overview of the Telegram plugin for Auto-GPT, covering the core functionality, error handling, asynchronous programming, and some additional considerations for refining the plugin. As we reach the conclusion of this walk-through, it's a good time to summarize key takeaways and suggest further steps for readers or developers looking to work with or build upon this plugin.

Here is a summary of the key takeaways:

- **Plugin structure**: The plugin comprises two main files: `__init__.py` and `telegram_chat.py`

  `__init__.py` is the entry point where Auto-GPT interacts with the plugin, while `telegram_chat.py` encapsulates the Telegram-specific logic

- **Initialization and configuration**: The `__init__` methods in both files are crucial for initializing and configuring the plugin, including setting up the Telegram bot credentials

- **User interaction**: The plugin allows for user input via Telegram and can send messages back to the user through methods such as `user_input`, `report`, `ask_user`, and `send_message`

- **Asynchronous programming**: Asynchronous methods and the use of the `asyncio` library enable non-blocking IO operations, improving the plugin's responsiveness and performance

- **Error handling**: Exception handling is employed throughout the code to catch and log errors, making the plugin more robust and resilient

Let us now look at the further steps that can be taken:

- **Explore the GitHub repository**: You are encouraged to explore the GitHub repository (`https://github.com/Significant-Gravitas/Auto-GPT-Plugins/tree/master/src/autogpt_plugins/telegram`) for the latest version of the plugin and to understand any updates or modifications made to the code.

- **Contribute to the project**: Developers interested in contributing can fork the repository, make their own improvements or additions, and submit pull requests. This collaborative approach can help enhance the plugin over time.

- **Implement suggested enhancements**: Implementing the suggested enhancements such as authorization checks, command handling, logging, environment variable management, code modularity, unit testing, and documentation could significantly improve the plugin's functionality and maintainability.

- **Experiment and customize**: Developers are encouraged to experiment with the plugin, customize it to fit their specific needs, and even extend it to incorporate additional features or integrations.

- **Learn and share**: Engaging with the community, learning from others, and sharing knowledge and experiences can be beneficial for everyone involved.

This walk-through aims to provide a thorough understanding of the Telegram plugin for Auto-GPT and offer a foundation for developers and for you all who are looking to delve deeper into plugin development for Auto-GPT. Through exploration, experimentation, and collaboration, the community can continue to build and improve upon this and other plugins, enhancing the capabilities and applications of Auto-GPT.

Summary

By dissecting the various methods and logic within the `__init__.py` and `telegram_chat.py` files, you will gain a thorough understanding of how the Telegram plugin is structured and operates. This step-by-step breakdown elucidates how Auto-GPT communicates with Telegram, handles user input, and sends messages or reports back to the user.

The complete code for this plugin, along with potential updates or modifications, can be found at `https://github.com/Significant-Gravitas/Auto-GPT-Plugins/tree/master/src/autogpt_plugins/telegram`. This repository is an excellent resource for those interested in exploring the plugin further or adapting it to their needs.

In this chapter, we provided a short introduction to plugins in Auto-GPT. We gave an overview of plugins, the different types of plugins and their use cases, how to use plugins effectively, and the process of building plugins. By leveraging plugins, you can extend Auto-GPT's functionality, tailor it to specific tasks or domains, and enhance its performance in various applications.

In the next chapter, we will delve into practical examples and case studies that demonstrate the power of plugins in real-world scenarios. If you have any specific requests or suggestions for modifications, please let me know!

5

Use Cases and Customization through Applying Auto-GPT to Your Projects

In the last chapter, we learned about plugins and how to customize them.

Building upon that foundation, this chapter transitions into practical applications, guiding you through the process of integrating Auto-GPT into your projects.

In this chapter, we will learn how to apply Auto-GPT to our projects by understanding the following topics:

- Setting up a chat assistant
- What to look out for
- Examples of customizations – the Telegram plugin and LLM plugins
- Diving deeper – the world of LLM plugins
- Custom characters and personalities of chats

Setting up a chat assistant

Just as every user has their own preferences and needs, so do I.

That is why I have created a Telegram plugin, which allows me to chat with my AI assistant via Telegram.

I made two versions: the official one, which is available on the Auto-GPT plugin repository, and a custom one, which has a little bit more of an individualized touch to it.

The official one is available in the Auto-GPT plugin repository and can be installed by cloning the repository into the `plugins` folder and just activating it in the `config` file. The rest of the setup is automated by the plugin.

The other one, which I called Sophie (to add more of an individualized touch), is available on my GitHub and can be installed by cloning the repository and activating it in the `config` file. It will remain in my repository fork at `https://github.com/Wladastic/Auto-GPT-Plugins` in case the Sophie plugin is not available in the official repository.

It is a command extension to Auto-GPT that lets the plugin chat with the user directly via Telegram (or let's any plugin chat with what they have implemented as the chat message service), and allows it to keep old conversations with a command that Auto-GPT can call to directly retrieve a summary of the conversation and the last few messages.

As Auto-GPT is an open source project and evolves with the community, I am sure that in the future there will be more plugins available that will allow you to customize your AI assistant even more, and this is also the reason why I will now focus on my own plugin, as I have a good amount of control over it and can make sure you as a reader can use it as well.

Research helper

Another example in which you could make use of Auto-GPT is as a research assistant.

This is something that could come in handy and make research a lot easier, such as when you need to remember something or look for something and you don't have the time or patience to search through Google. It might be useful to have automation to do that for you.

By using a research helper plugin for Auto-GPT, you could input a research query and Auto-GPT will then search for it and give you a summary.

For example, you could input `What is the average GDP of South Korea?` and Auto-GPT will search for it and give you the most relevant results.

Speech assistant

Another potential use case would be to use Auto-GPT as a speech assistant. This could be practical if you do not have the time or the energy to look up something on the web or you are just feeling lazy.

When using Auto-GPT as a speech assistant, you can input the queries by speech, and Auto-GPT will then search for them and give you the most relevant results. For example, you could ask *What is the GDP of South Korea?* and Auto-GPT will find and provide you with the most relevant results.

Custom characters and personalities of chats

Finally, Auto-GPT can also be used to create custom characters and personalities that can appear in your chat conversation. This could be useful if you want to create a chatbot that is more approachable and conversational with the user. To do this, Auto-GPT can be used to create custom chatbot personalities and customized conversation styles, all with the same look and feel.

For example, the look and feel of the conversation could be that of a sci-fi movie or a novel. The character could also have custom emotions or a custom style of responding to the users.

All this can be done right in the Auto-GPT `config` file in the chatbot section or in the plugin. This will give you the choice to customize your own characters and personalities to add that individualized touch to your project.

What to look out for

When creating customizations with Auto-GPT, it is important to watch out for features that may be easy to overlook. For example, when creating a chatbot, it is important to watch out for features such as forgetting what was said in an earlier conversation or not being able to understand a user's commands. Also, when creating custom characters or personalities, it is important to watch out for features such as emotions and conversation patterns that may seem unnatural. This can be especially important for ensuring a natural conversation between the user and the chatbot.

Overall, when using Auto-GPT to customize a project, it is best to make sure to think about all the features and potential issues to ensure that the experience your user has with your project is perfect.

However, only relying on the base features is not enough, not to mention boring, so let's explore our options.

Examples of customizations – the Telegram plugin and LLM plugins

A few examples of customizations with Auto-GPT include the Telegram plugin, LLM plugins (explained ahead in the chapter), and custom characters and personalities of chats.

The Telegram plugin allows users to chat with their AI assistant directly via Telegram, and the LLM plugins allow users to customize their AI assistant's language, learning models, and memory.

Custom characters and personalities of chats allow users to create custom personalities and conversation patterns with the same look and feel. This can be done right in the Auto-GPT `config` file in the chatbot section or in the plugin.

These are just a few of the many potential customizations available with Auto-GPT. As Auto-GPT is an open source project, the possibilities are limitless – it all depends on what you want to achieve and how you want to customize your project.

The vast landscape of AI-driven applications is punctuated by the need for customization, and Auto-GPT rises to the occasion, offering a plethora of customization avenues. Among the myriad plugins and features, the Telegram plugin, LLM plugins, and custom chat personalities stand out as sterling examples of what is achievable.

One of those is my Telegram plugin, which, in the newest version of Auto-GPT, has now even been added to the base code.

Telegram plugin – bridging conversations

Telegram provides a seamless integration. The Telegram plugin is emblematic of the seamless fusion between messaging platforms and AI. This integration transforms the ubiquitous messaging app, Telegram, into a conduit for real-time, intelligent conversations with an AI assistant. Imagine having a pocket researcher, advisor, entertainer, and assistant all rolled into one in your Telegram chats.

Under the hood, the Telegram plugin uses the powerful APIs of both Telegram and Auto-GPT. Once activated in the Auto-GPT configuration, it listens for incoming messages on Telegram, processes them in real time through Auto-GPT, and returns AI-generated responses, fostering dynamic conversations.

It has the following benefits:

- **Ubiquity**: With Telegram available across devices, your AI assistant is always within reach
- **Security**: Telegram's end-to-end encryption ensures that conversations remain private
- **Multimedia capabilities**: Exploit Telegram's multimedia features, allowing your AI to process and respond to images, voice notes, and more
- **LLM plugins**: Tailor an intelligent mind

Language is not just a means of communication; it is an intricate tapestry woven from cultural nuances, historical contexts, and ever-evolving semantics. Auto-GPT, with its innovative capabilities, recognizes this complexity and offers **language learning models** (**LLMs**) to navigate this vast linguistic landscape.

What are LLMs?

At their core, LLMs form the very essence of Auto-GPT's linguistic prowess. They serve as the AI's neural framework, determining its comprehension, learning mechanisms, and response patterns across diverse languages and contexts. Think of LLMs as the sophisticated "brain" behind Auto-GPT, processing language intricacies with finesse.

While rudimentary AI models might translate languages verbatim, LLMs delve deeper. They capture idiomatic expressions, colloquialisms, and cultural nuances, ensuring that the AI's interactions remain authentic and contextually relevant.

LLMs are not static; they evolve. As they process more data, they refine their understanding, adapt to linguistic shifts, and enhance their proficiency, mirroring the fluid nature of languages.

Now that we have explored what LLMs are, we can look at the default LLM we use and ask: *Why do we only use one tool if there are millions out there?*

Let's answer that question.

Diving deeper – the world of LLM plugins

While the foundational LLMs offer expansive linguistic capabilities, the real magic lies in customization. LLM plugins function as specialized extensions, allowing users to mold Auto-GPT's language capabilities to fit precise requirements.

A multitude of possibilities

Whether you are a business aiming to offer customer support in a regional dialect, a researcher seeking insights from niche linguistic datasets, or a storyteller wanting to craft narratives in multiple languages, LLM plugins supply the tools to make these visions a reality.

Key features of LLM plugins

While most functions can be fulfilled by the default OpenAI GPT models, some are better fulfilled by custom LLMs. To use those, we will either have to modify Auto-GPT or add them with plugins.

The global and the local

In an increasingly globalized world, communication transcends borders. LLM plugins empower Auto-GPT to converse not just in widely spoken languages but also in regional dialects and lesser-known tongues.

By supporting diverse languages, LLM plugins ensure that the AI tool's interactions are still genuine, respectful of cultural nuances, and free from generic translations.

Domain specialization – ability at your fingertips

Language is vast, but ability often lies in niches. Whether you are looking at the technical jargon of quantum physics, the intricate terminologies of law, or the nuanced language of literature, LLM plugins can be tailored to master specific domains.

Real-world implications

Imagine a med-tech start-up harnessing an LLM plugin that specializes in medical terminologies, easing accurate patient interactions, or a legal firm employing an LLM-enhanced AI to sift through case laws, extracting relevant insights.

Memory management – balancing recall and privacy

Just like humans, an AI's effectiveness hinges on its memory. However, while recall is vital to context, privacy is paramount. LLM plugins strike this balance, allowing users to define how the AI keeps or forgets interactions.

For instance, in customer support scenarios, keeping the context of past interactions can foster continuity, but purging personal data post-interaction is crucial to user trust and compliance with data protection regulations.

While GPT is very powerful, it lacks customizability unless we are ready to wait and trust OpenAI to improve its models and to also become reliable enough that we can safely use them. There is also the option to use other LLMs as our core, and maybe even better ones.

The future of LLM plugins

As the realms of AI and linguistics converge, the potential for LLM plugins is boundless. With continuous research, community contributions, and real-world feedback, these plugins will continually redefine the boundaries of what is achievable in AI-driven linguistic interactions.

The open source ethos of Auto-GPT ensures that LLM plugins benefit from global intelligence. Developers from diverse linguistic and cultural backgrounds contribute to this ecosystem, enriching LLMs with their unique perspectives.

In an era when communication is paramount, LLM plugins stand as sentinels, ensuring that language, in all its richness and diversity, is celebrated, understood, and used effectively.

Redefining interactions

Beyond mere utility, there is an art to crafting AI interactions. With Auto-GPT, you are not just configuring a chatbot; you are breathing life into a digital entity, defining its personality, demeanor, and conversational style.

The creation process

Whether you envision a chatbot with the wit of Oscar Wilde, the wisdom of Confucius, or the charisma of Oprah Winfrey, Auto-GPT's customization capabilities make it achievable. Delve into the config files, tinker with conversation patterns, set emotional responses, and curate a unique digital persona.

Applications

The following are a few applications of LLM plugins:

- **Customer support**: Design empathetic chatbots that resonate with users, offering support with a human touch
- **Entertainment**: Create fictional characters for interactive stories or role-playing games
- **Education**: Craft tutor personas tailored to different learning styles and subjects

Unleashing potential – the open source advantage

Auto-GPT's open source nature is its trump card. It not only democratizes AI but also fosters a vibrant community of developers, researchers, and enthusiasts. This collective intelligence continuously expands the horizons of what is possible.

The community edge

With a global community contributing to its ecosystem, Auto-GPT receives help from diverse perspectives, novel ideas, and innovative solutions. This ensures that the platform stays dynamic, evolving with emerging needs and trends.

Your canvas

As you embark on your customization journey, remember that Auto-GPT is akin to a canvas. While it offers the tools and the palette, the masterpiece you create is limited only by your imagination. However, keep in mind that the more complicated your goal becomes, the more work you will have to put into improving Auto-GPT to actually reach that goal.

For most of our tasks, we will need memory functions; otherwise, Auto-GPT becomes very forgetful. Therefore, we also need embeddings to make as much data accessible as possible.

Custom embedding and memory plugins – the evolutionary aspects of Auto-GPT

Auto-GPT is architected with a vision that extends beyond traditional AI models. One of its revolutionary design choices is treating the GPT not as a monolithic core but rather as a foundational plugin. This design philosophy ensures that while GPT offers a robust starting point, it is merely a part, one that can be replaced or augmented. Such a modular approach positions Auto-GPT for future advancements, ensuring that it stays adaptable, extensible, and perennially relevant. This section delves deep into two pivotal dimensions of this modularity: custom embedding plugins and custom memory plugins.

Having embeddings checked as a topic, we can now move a bit toward the base of Auto-GPT, which is the LLM we use as our thinking core, so to speak. Although GPT-4 from OpenAI is currently the go-to LLM, Auto-GPT presents some very user-friendly ideas.

The GPT as a base plugin – the first building block for inspiration

Nothing shows more openness to customization than taking the first step toward it. Because Auto-GPT itself is designed to be as modular as possible, it was decided that it would treat as many functions that Auto-GPT provides as a module or even a plugin. Without an LLM, Auto-GPT is nothing but a servant that cannot decide to do anything. Based on that, OpenAI's **generative pre-trained transformer (GPT)** is the base plugin.

Traditional AI models are designed with static architectures, where core components, once integrated, remain immutable. Auto-GPT challenges this norm. By treating GPT as a base plugin, it sets the stage for endless customizations and improvements.

Visualize Auto-GPT as a dynamic mosaic, where each tile (or plugin) contributes to the whole picture. The GPT tile, while significant, is just one among many. It can be replaced, refined, or complemented by other tiles to create new patterns and functionalities.

This approach ensures that Auto-GPT remains unfettered by the limitations of any specific model or technology. As AI evolves, newer, more advanced models can be integrated seamlessly, ensuring that Auto-GPT is still at the forefront of technological advancements.

Custom embedding plugins – refining the language of AI

In the realm of **natural language processing** (**NLP**), embeddings are akin to the DNA of language. They translate words and phrases into numerical vectors, capturing semantic nuances, relationships, and contextual meanings.

Why custom embeddings?

While GPT offers a comprehensive embedding mechanism, specific applications may demand nuanced linguistic representations. Whether it is capturing the subtleties of regional dialects, the jargon of niche domains, or the evolving lexicon of internet slang, custom embedding plugins supply the tools to tailor these representations.

We have three such custom embeddings:

- **Domain-specific embeddings**: Imagine a pharmaceutical company striving for correct drug nomenclature. A custom embedding can be designed to understand the vast array of drug names, their variants, and their relationships, ensuring precision in AI interactions.

- **Cultural and regional embeddings**: For a global brand aiming to resonate with diverse audiences, embeddings can be tailored to understand idioms, colloquialisms, and cultural references specific to various regions.

- **Evolving embeddings**: Language is fluid, and constantly evolving. Custom embedding plugins can be designed to be dynamic, adapting to linguistic shifts and incorporating new terminologies and phrases in real time.

Custom memory plugins – the art of recollection and forgetting

One of the main functions that Auto-GPT needs to work is memory. If we do not provide it with any memory, it is nothing more than a ChatGPT clone that can do one extra step.

The role of memory in AI

Memory in AI mirrors its significance to human cognition. It decides how an AI recalls past interactions, learns from them, and uses them to inform future decisions.

Why custom memory?

While GPT offers a robust memory mechanism, there's no one-size-fits-all solution. Different applications demand varied memory behaviors. Some require long-term retention for continuity, while others prioritize short-term memory for privacy.

Crafting custom memory mechanisms

One-shot mechanisms, where we gather the whole memory and give it to the LLM, can be easy to implement but can introduce a huge load of new contexts that do not necessarily add any value to the current context that Auto-GPT is working in. For example, telling it about a coding project it worked on with you while you're just asking it about the current time is clearly overkill.

Adaptive retention

Consider a customer support chatbot. While it needs to recall past interactions for continuity, it must also forget personal data post-interaction to ensure privacy. Custom memory plugins can strike this balance, adapting retention based on context.

Learning and unlearning

In dynamic domains such as stock markets and healthcare, where outdated information can be detrimental, custom memory plugins can be designed to periodically unlearn outdated knowledge, ensuring the AI's insights are still current.

Contextual memory

For applications such as storytelling and role-playing games, memory plugins can be designed to remember plot points, character arcs, and user choices, ensuring continuity and immersion.

In conclusion – the infinite horizon of customization

Auto-GPT, with its foundational philosophy of modularity, opens up a realm of possibilities. Treating GPT as a base plugin and offering avenues for custom embedding and memory plugins ensures that AI developers and innovators are limited only by their imaginations. As the AI landscape evolves, Auto-GPT stands poised to adapt, evolve, and lead, offering tools that are not just technologically advanced but also profoundly customizable.

This exploration into custom embedding and memory plugins underscores Auto-GPT's commitment to flexibility, adaptability, and forward-thinking design, ensuring it remains a torchbearer in the ever-evolving domain of AI.

Custom characters and personalities of chats

Creating custom characters and personalities with Auto-GPT is a wonderful way to add that individualized touch to your project. To create a custom character, you will need to think about personality and conversation patterns. You should also consider the feelings and emotions that this character should have in order to have a natural conversation with the user.

You will also need to think about the design of the character, as this will help you to create a more consistent look and feel for the conversation.

Finally, you also need to consider how the character will react to different commands and situations. This will help to ensure that the user has an enjoyable and natural conversation with the character.

Overall, with Auto-GPT, you can customize anything you can think of – it is all up to you.

Summary

In this chapter, we learned how to apply Auto-GPT to our projects and how to customize characters for a better user experience.

We learned how to set up an AI chat assistant and a research helper using Auto-GPT, as well as how to use it as a speech assistant. We also discussed the importance of creating custom characters to create a more natural conversation.

Finally, we looked at what to look out for when customizing a project with Auto-GPT and reviewed a few examples of customizations that can be done.

Overall, Auto-GPT provides a great deal of customization and can be used to create many unique projects. With its open source nature, the possibilities are virtually limitless and it's a powerful tool for all kinds of projects.

In the next chapter, we will explore the advanced setup in Docker, outside of the already existing Docker image run that might be much easier, but as we talk about customization, this is more than necessary.

6

Scaling Auto-GPT for Enterprise-Level Projects with Docker and Advanced Setup

In *Chapter 2,* we covered how to start Auto-GPT with Docker; now, we will dive deeper into the utilization of Docker.

Docker has become a dynamic tool in the realm of software development, especially in the management and distribution of complex applications such as Auto-GPT. As such, this chapter aims to provide you with a comprehensive understanding of how Auto-GPT utilizes Docker. Using Docker will make sure we are always on the same page. Also, you can read about several cases, such as this GitHub issue I reported which has an early issue number `https://github.com/Significant-Gravitas/AutoGPT/issues/666`, where my Auto-GPT Agent managed to break out of its own boundaries using an exploit it found by itself.

We will also explore the continuous mode functionality in Auto-GPT and discuss its implications. The power of Auto-GPT is not only in its ability to generate creative and coherent content but also in its capability to operate autonomously.

This chapter will also delve into one of Auto-GPT's key features: continuous mode. We will explore what it is, its potential applications, and the precautions required when using it.

Continuous mode in Auto-GPT allows the program to run without requiring user approval at every step. This means that it can generate content, perform tasks, and even make decisions independently. This feature is particularly useful for automating tasks that would otherwise require constant human intervention. However, as with any powerful tool, it is essential to use it responsibly.

This chapter covers the following topics:

- An overview of how Auto-GPT utilizes Docker
- Fixing potential loopholes or bugs
- Example run scripts
- What is continuous mode?
- Known use cases of continuous mode
- Safeguards and best practices
- Potential risks and how to mitigate them
- Gracefully stopping a continuous process

An overview of how AutoGPT utilizes Docker

Docker operates on the principle of containerization – a streamlined and isolated way to run applications securely and efficiently. Consider these containers as self-sufficient units housing all the components required to run an application. Auto-GPT's integration with Docker empowers users to bypass the strenuous process of setting up environments manually, turning Auto-GPT's focus towards customizing AI experiences instead.

Essentially, Docker encapsulates Auto-GPT's computational environment. The Docker container for Auto-GPT includes the Python environment, the necessary libraries, as well as the application itself. The Dockerfile, located in the root directory of Auto-GPT, holds the instructions for Docker to build this image.

So, how it works is that Docker creates an isolated environment, or *container*, where Auto-GPT runs. Every interaction between Auto-GPT and your system goes through Docker, which interprets these interactions and ensures they're safe and compatible with the environment within the container. Isolation also means that if anything goes wrong inside the container, your system remains unaffected.

The Dockerfile for this often just contains a few commands, such as which environment is supposed to be loaded, which is `python:3.10-slim` on top of poetry in the case of Auto-GPT.

Let's now see how this integration works.

Understanding Auto-GPT's integration with Docker

The integration of Docker with Auto-GPT provides several advantages. The encapsulation ensures that each user experiences precisely the same computational environment, thereby reducing disparities that could arise due to different operating systems or Python distributions. This encapsulation also permits seamless sharing and easy version control, and eliminates the *It works on my machine* problem!

Auto-GPT had a memory database with SQLite running that contained the actions and messages of prior runs. This was first augmented with a few vector database solutions at the beginning of 2023, but was completely dropped by the middle of 2023 because some of them did not work with most users. Since then, a VectorDB or any memory apart from saving into a file locally has not been implemented because supporting too many solutions was too much work. Instead, the memory is held in a JSON file locally, whose position may vary in each release version of Auto-GPT. Using Docker was a necessary step here, although many users then (including me) complained that Auto-GPT was less supportive of local running instances. Fortunately, it changed later, as the memory system is again a very basic JSON file now, which is better than nothing.

Having delved into why Auto-GPT supports Docker, we may now move forward to how to start our Auto-GPT Docker instance.

Starting a Docker instance

While this may sound complex, running a Docker container is remarkably straightforward. Firstly, if you haven't already, you need to make sure Docker is installed on your system. Several guides can assist you with this process, such as Docker's official installation instructions (`https://docs.docker.com/get-docker/`), which can guide you through the process for Windows, MacOS, and various Linux distributions.

To run an instance of Auto-GPT in a Docker container, you'll first need to build the Docker image using the Dockerfile. Here's a quick rundown:

1. Navigate to the directory that contains the Dockerfile (it should be the root directory of Auto-GPT): `cd path/to/Auto-GPT`.

2. Build the Docker image with `docker build -t auto-gpt ..` (make sure it ends with a space and dot; it may not be clearly visible).

    The easy way to run the instance is to execute the following:

    ```
    docker compose run --rm auto-gpt --gpt3only --continuous
    ```

3. If you dare, you can also build and run it with vanilla Docker commands:

    ```
    docker build -t auto-gpt .
    docker run -it --env-file=.env -v $PWD:/app auto-gpt
    docker run -it --env-file=.env -v $PWD:/app --rm auto-gpt
    --gpt3only --continuous
    ```

If you have followed all the steps correctly, you should be greeted with Auto-GPT's chat input. If not, we will go through the common ways it may not work. Also, you can refer to the previously mentioned Docker documentation if the functionality may have changed.

Although Docker is often easy to just install and go, we may still run into some problems.

Fixing potential loopholes or bugs

To use Docker on a PC, you have to make sure virtualization is activated.

On macOS, you do not have to change anything. Virtualization is already working there; you only have to download and install Docker Desktop.

As each motherboard manufacturer has different BIOS settings, the way you activate it may vary, but normally you would go to the BIOS by restarting your machine and pressing the *Setup/BIOS* button that is displayed. For me, it's *F2* or *DEL*. If no button is displayed, you will have to do some research.

Depending on which CPU manufacturer you have, the setting may be less obvious. For example, Intel has a setting named **Intel Virtualization Technology**, while AMD has **AMD-V**. Once you have found it, make sure to enable it and save the change.

It's important to mention that working with Docker can sometimes lead to bugs or issues that are not apparent at first. An example of such is the notorious **Docker space** error. Docker uses a part of your system storage to store its images and containers, and sometimes this space can get filled up without any clear warning. This can result in the software abruptly stopping or not working as expected. To resolve this, you can periodically clear up space by using the `docker system prune` command.

Despite Docker's advantages, it isn't without potential challenges. One common issue involves insufficient memory allocation. When you start Docker, it reserves a certain amount of memory which, if depleted, can cause unexpected behavior or crashes. To resolve this, try to increase Docker's memory allocation in Docker's settings.

Another common issue is when Docker fails to start a container because it claims the selected port is already in use. Even if you're certain it isn't, a misbehaving application could be invisibly binding to that port. In this case, rebooting the system or manually killing the process occupying the port can help.

I personally recommend either the Docker Desktop app on Windows and Mac, or for Linux, I use Portainer. This is a web-based Docker manager application that is just as powerful as the official counterpart.

You can install and run it by executing these commands:

1. First, create the volume that the Portainer Server will use to store its database:

    ```
    docker volume create portainer_data
    ```

2. Then, download and install the Portainer Server container. Keep in mind that this should all be on one line:

    ```
    docker run -d -p 8000:8000 -p 9443:9443 --name portainer
    --restart=always -v /var/run/docker.sock:/var/run/docker.sock -v
    portainer_data:/data portainer/portainer-ce:latest
    ```

    Once it starts, you will see a password on your console that you will need to log in to the UI. Just go to `localhost:8000` or `https://localhost:9443`; your username is `admin`.

    If no password was displayed in the console, you will be asked to set a new password at this stage.

Docker also creates intermediate images during the build process. These images can take up a substantial amount of disk space. Check your Docker system by running `docker system df`, and clean up unneeded images with the `docker system prune` command.

Sometimes, it is not the project's fault; sometimes, the software we use may experience some issues. Docker is very commonly used, but it can also cause some problems.

Identifying and fixing potential issues related to Docker

Utilizing Docker alongside Auto-GPT can be a rewarding experience, but like all tech integrations, it comes with a set of challenges.

Here's an in-depth guide to tackling these challenges:

- Docker daemon not running:

  - **Symptom**: Errors suggesting Docker isn't operational.

  - **Solution**: Here are the solutions for each OS:

    - **On Linux**: Open the terminal and type `sudo systemctl start docker` to initiate the Docker service.

    - **On Windows**: Navigate to the **Start** menu, locate **Docker Desktop**, and run it. It will initiate the Docker service.

    - **On macOS**: Launch the Docker application from the `Applications` folder. If Docker is running correctly, you'll see the Docker whale icon in your top menu bar.

- Auto-GPT packages or modules missing:

  - **Symptom**: Exceptions/errors saying a module has not been found.

  - **Solution**: We can use these solutions:

    - **Access the container**: To interact directly with a running Docker container, use the following command:

      ```
      docker exec -it <container_name_or_id> /bin/bash
      ```

      This command will open a bash shell inside the container, allowing you to execute commands directly within the container's environment.

    - **Navigate to Auto-GPT directory**: Once you're inside the container, navigate to the directory where Auto-GPT is installed. This location might vary based on your setup, but as an example, it might be something like `cd /app/`.

- **Install missing modules**: Within the Auto-GPT directory, you'll likely find a `requirements.txt` file that lists all necessary packages. Use `pip` to install these requirements:

- `python -m pip install -r requirements.txt`.

- This command will ensure all listed packages in the `requirements.txt` file are installed within the container's environment. If there are any additional modules or packages you know are missing, you can install them individually using `pip` as well.

- **Exit the container**: Once you've successfully installed the required packages, you can exit the container's bash shell by typing `exit`.

- Remember, Docker containers are isolated environments. Any changes you make inside a container (such as installing packages) won't affect your host system. However, these changes will be lost if the container is removed. To make persistent changes, you might consider creating a new Docker image or using Docker volumes.

- Insufficient memory or disk space: Docker may sometimes use up too much space, or maybe you just do not have enough storage available. Docker may just shut down and not even tell you what was wrong:

  - **Symptom**: Containers exit prematurely or fail to initiate.

  - **Solutions**: Here are some solutions:

    - **Check space utilization**: Run docker system df to get a snapshot of your Docker's disk usage.

    - **Clean Up**: Use `docker system prune -a` to remove unused data. Note: This will delete unused containers, networks, and images. Always backup essential data.

    - **Adjust resources (Windows/macOS)**: Open **Docker Desktop | Settings | Resources** to modify memory or disk allocations.

- Networking issues: Sometimes, the port forwarding to and from Docker containers may be broken. For example, when we want to open the frontend website of our Auto-GPT instance, it may be that we cannot access it:

  - **Symptom**: Connectivity problems between your container and external services

  - **Solutions**: Here are some solutions:

    - **Inspect ports**: With `docker port <container_name>`, you can check which ports are active

    - **Modify port bindings**: When initiating a container, use the -p flag, such as `docker run -p 4000:80 <image_name>`, to bind container ports to host ports

- Version conflicts:

  - **Symptom**: Unforeseen errors possibly due to outdated software.

  - **Solutions**: For Docker updates, you can try the following:

    - **On Linux**: Depending on your distribution, use `sudo apt-get update && sudo apt-get upgrade docker-ce` or similar commands to update Docker.

      - **Windows/macOS**: Docker Desktop will notify you of available updates. Ensure you keep it up to date.

    - For Auto-GPT updates, regularly check the official Auto-GPT repository or documentation for the latest versions and update instructions.

- Container isolation:

  - **Symptom**: Applications inside a container can't access host services or files

  - **Solution**: Here are a couple of solutions:

    - **Mount directories**: When starting a container, use `-v` to mount host directories to the container, such as `docker run -v /host/directory:/container/directory <image_name>`

    - **Network configuration**: Use `docker network inspect <network_name>` to check the network configuration and ensure containers are correctly connected

- Logs and diagnostics:

  - **Symptom**: Ambiguous errors or container behaviors.

  - **Solution**: Here is a solution:

    - **Access logs**: The `docker logs <container_name>` command will display the logs of your container. These logs are invaluable for diagnosing issues. If an error occurs, it's likely detailed in these logs.

Remember, the keys to troubleshooting are patience and systematic exploration. With these detailed steps, you'll be well equipped to handle most Docker-related challenges that come your way when working with Auto-GPT.

Sometimes the issues we experience are deeper, and for those, we need to access the container that Auto-GPT is running on.

To get an idea of how else you may start Auto-GPT using Docker, I have made a few examples.

Example run scripts

To familiarize yourself with run scripts, you can refer to the ones listed here:

- To run an interactive shell in the Docker container (useful for debugging), use the following command:

```
docker run -it --entrypoint /bin/bash auto-gpt
```

This command starts an interactive shell within the Docker container, allowing you to directly access and debug Auto-GPT.

- To run Auto-GPT on a different port, try this:

```
docker run -p 4000:5000 auto-gpt
```

By specifying the port mapping with the -p flag, this command runs Auto-GPT on port 4000 of your host machine, while internally, it listens on port 5000 within the Docker container.

- To forward all traffic from port 80 to Auto-GPT (this requires administrator/sudo privileges), run the following:

```
sudo docker run -p 80:5000 auto-gpt
```

This command enables port forwarding, allowing all traffic from port 80 on your host machine to be redirected to Auto-GPT running on port 5000 within the Docker container. Note that it requires administrative or sudo privileges.

By utilizing these run scripts, you can enhance your efficiency and productivity when working with Auto-GPT in the Docker environment. Feel free to experiment with different port mappings or explore the interactive shell for debugging purposes.

As we move forward, brace yourself for an exploration of the intriguing world of Auto-GPT's continuous mode. We will delve deep into its mechanics, understanding the potential consequences and benefits it brings to the table, and as the chapters unfold, we will also touch upon integrating different LLM models, making the most of prompts, and much more. The journey with Auto-GPT is about to get even more exciting!

Let's move onto continuous mode and its consequences.

What is continuous mode?

We will start by explaining what continuous mode is. It allows Auto-GPT to run without requiring user approval at each step. This enables automation without constant human oversight.

We will then look at some known use cases and examples of continuous mode, such as automating research tasks, content generation, and code compilation. While continuous mode is powerful, note that caution should be exercised given the lack of human verification.

Next, we will provide tips for using continuous mode safely and effectively. This includes setting clearly defined goals, constraints, and limits in the configuration to restrict unwanted behavior. Monitoring logs and outputs is also advised.

Furthermore, we will discuss potential risks, such as generating falsities, getting stuck in loops, and cost overruns. Mitigation strategies such as kill switches, usage limits, and sandboxing will be suggested.

Lastly, we will explain how to gracefully stop a continuous process, such as using keyboard interrupts or the `task_complete` command. We will also explore active research into enhancements such as pause commands and timed takeovers.

Continuous mode allows Auto-GPT to operate autonomously without requiring user approval at each step. By keeping the core loop running uninterrupted, it enables the automation of tasks without constant human oversight. However, caution is advised. When used judiciously with appropriate safeguards, continuous mode can boost efficiency. When deployed carelessly, it can have regrettable consequences. Finding the right balance requires an understanding of its capabilities and risks. This part of the chapter aims to provide that holistic perspective.

Known use cases of continuous mode

My own favorite use case is a personal assistant that only reports or asks when input is needed. While most AI chatbots currently work one step at a time, I prefer Auto-GPT to run continuously, and I wrote my Sophie-Plugin so that Auto-GPT only texts me when it either wants to report something and keeps going (it then adds . . . (it's just three dots; I did not add any other symbols, as it may create confusion) at the end of the message), or it asks me to make a certain decision or provide feedback.

This way, the assistants' use case is broader, and you can also give it multiple tasks and tell it to get back to you once they are done, in combination with other plugins, such as email plugins, discord plugins, and SSH terminal plugins. Your AI is even capable of managing communication for you, which could be risky but is at the same time very cool. Auto-GPT once even negotiated competitive salaries for me for a coding project it was supposed to do as it reached out to companies to test the code. On a side note, that was GPT 3.5 before it was Turbo. The new models are not as bold or aware of what the context is anymore, which could maybe be achieved with alternative LLM models, which were used in my mini-autogpt project (`https://github.com/Wladastic/mini_autogpt`), where I lined up several LLM models and tested their congruence. As OpenAI seemingly tries to minimize the cost of running its new LLMs, it has become less and less sure about what they do and they are either ambivalent or repetitive. Also, they feel like they do not really listen to you and stick to what they said previously.

Automating research and analysis

One of the most compelling applications of Auto-GPT's continuous mode is in the field of research and analysis. By activating continuous mode, Auto-GPT can endlessly mine information from multiple sources, such as academic journals, news feeds, and social media platforms. Market researchers can deploy this feature to keep an ongoing tab on consumer sentiments, emerging trends, and competitor strategies. The feature is also highly beneficial for academic research, as it can scan newly published papers and add new insights in real time to an ongoing study.

Streamlining content creation

Content creators and marketing departments can significantly benefit from Auto-GPT's continuous mode. The AI can be programmed to scan trending topics and produce relevant articles, blog posts, or social media updates without human intervention. Furthermore, because the model can generate content in various styles and formats, it allows a diversified content strategy. Imagine running a news website where the AI continually updates content based on emerging global events, freeing up human editors to focus on more complex tasks, such as investigative reporting or opinion pieces.

Supercharging code compilation

Developers often find themselves waiting for code to compile, which interrupts the flow of work. Auto-GPT in continuous mode can manage these mundane tasks. It can be set up to automatically compile code, run test cases, and even push updates to a repository. This enables a more seamless development process and ensures that the latest changes are always integrated and tested. By offloading routine tasks to Auto-GPT, developers can focus on problem-solving and innovation.

Always-on customer support

Customer support is an area where Auto-GPT's continuous mode can be transformative. By enabling this feature, the chatbot can handle an infinite number of queries and concerns without requiring human intervention. This makes for a 24/7 customer support system that can handle most issues and only escalates complex or sensitive matters to human agents. This not only improves customer satisfaction but also significantly reduces the operational costs associated with customer service.

Before we dive into using continuous mode, let's first make sure we learn some best practices.

Safeguards and best practices

While the capabilities of Auto-GPT's continuous mode are impressive, it is essential to approach its deployment with caution. Here are some measures you can take:

- **Enable confirmation prompts**: For commands that could potentially incur costs or are irreversible, enable confirmation prompts. For example, if you set up Auto-GPT to handle emails, a confirmation prompt sent to the user before sending could prevent unwanted communications.

- **Use allowlists and blocklists**: Restrict the model's capabilities by using allowlists for approved actions and blocklists for prohibited ones. For example, you could use an allowlist to specify which external databases the AI can access for information.

- **Gradual resource scaling**: Begin with conservative computational and financial limits. As you observe the system's behavior and performance, you can slowly relax these constraints. This minimizes the risk of runaway costs or overutilization of resources.

- **Sandboxed testing**: Before rolling out the system in a live environment, it is advisable to test it in a sandboxed or isolated setting. This allows you to identify and correct any bugs or vulnerabilities without affecting real-world operations.

We have discussed ways to ensure good behavior. Now, we can assess how human monitoring through user prompts relates to general human oversight.

Regular monitoring and human oversight

While continuous mode aims to minimize human intervention, periodic reviews and adjustments are necessary. Monitor logs to detect any anomalies or irregular behavior and have a plan in place for immediate human takeover in high-risk scenarios.

However, human oversight is still indispensable. Adjust configurations based on feedback, establish approval workflows for high-risk scenarios, and default to human takeover when unsure.

If you want to fully trust GPT to make all the decisions itself and are confident that it might run perfectly, let's discuss the downside of having GPT make decisions for all actions and shake up your trust towards it a little.

Potential risks and how to mitigate them

As GPT is the core of our Auto-GPT application, it can and will at some point make absurd decisions.

For example, I let Auto-GPT run for a while with instructions to write me a web-based browser game whose genre is RPG that should be running with `Three.js`, and I got this:

- First, Auto-GPT created some empty files of the code it wanted to write but only added TODOs in them that indicate that the code has to be finished.

- Then it wrote some actual unit tests that might even partially work, but it got so confident that it finished the whole project and tested it. Then, it started searching Google to find companies to test the game and even wrote an email to one of them negotiating a price of only $100 per day as I set that as the budget for the project.

The way it wrote emails was scary and interesting at the same time. It found a coding website that just runs code you enter into a box, which also allowed URL params that accept code that is to be executed. Auto-GPT created a URL with the code it wrote as the parameters that sent an email to some company it found on Google that offered cheap QA testing.

It also managed to receive the email that was sent back with an answer for the price asked by the company employee, although to this day I have no idea how it created that email account because it was done by an agent that it created, and back then, they did not log what they did.

I was sitting on the couch and listening to the *speak* outputs of Auto-GPT. I set it to use a female voice for the main agent and for any other agent, it was a male voice, so I just thought it sounded like a conversation in an office until the male voice said *finished negotiating the daily price for testing with ****, reporting back from duty.*

Auto-GPT has changed since then. It is a bit stricter with the agents. Back then, the agents did whatever they wanted until they decided they were done with the task.

Despite precautions, preparedness to forcibly stop is crucial. But balance is key: avoid prematurely halting long-running tasks.

If we do run into a situation where we want to stop Auto-GPT from going further, but we do not want to break it, we will explore how to avoid abruptly killing it.

Gracefully stopping a continuous process

Sometimes, we may observe Auto-GPT behaving totally incorrectly and performing very wrong actions, or we just get a feeling that it is not doing anything productive, or it might even get so off track that it may run into a loop of researching a topic over and over without ever finalizing its search.

For example, when you see in the logs that Auto-GPT is stuck in a loop of doing the same few tasks over and over, you may want to stop it to prevent high running costs for using the API.

Keyboard shortcuts such as *Ctrl + C* send immediate cancellation requests.

Make sure to have the terminal in which Auto-GPT is running active by clicking on it if you are not sure.

If this does not work, for example, when Auto-GPT is *thinking* or a plugin is doing something and ignores the key interruption, you can always use a Process Explorer tool such as **Windows Task Manager**, **Mac OS Activities**, or **htop/top** on Linux to find the Python process of Auto-GPT and kill it manually.

When your Auto-GPT instance is running on Docker and continuous mode is enabled, always make sure that you have at least the Docker Desktop app open so you can shut down the Docker container of Auto-GPT if you see that it does something unexpected and does not react to *Ctrl + C*.

If you are away from your computer and only communicate to Auto-GPT over Telegram or Discord, for example, make sure to have a failsafe that always works to prevent Auto-GPT from becoming destructive.

When Auto-GPT decides that it has no other tasks, it also executes the `task_complete` command, which cleanly finishes ongoing tasks before stopping.

Pause/resume functionality is being explored to delay Auto-GPT actions without disrupting them. These delays could enable periodic human reviews to make sure no odd behavior starts happening.

In this section, we've examined continuous mode in Auto-GPT, a feature that allows the AI to operate autonomously without requiring user input at every step. This mode is very useful for automating tasks such as research, content creation, code compilation, and customer support, providing significant efficiency gains.

We discussed practical use cases, such as personal assistants and automated research tools, highlighting how continuous mode can streamline various processes. However, using this mode requires careful planning and safeguards. Best practices include enabling confirmation prompts, using allowlists and blocklists, starting with conservative limits, and conducting sandbox testing.

Regular monitoring and human oversight remain essential, even with continuous mode enabled. It's important to review logs and have a plan for immediate intervention if necessary. Potential risks, such as generating false information or getting stuck in loops, can be mitigated with strategies such as kill switches and usage limits.

Lastly, we covered how to stop a continuous process gracefully and the ongoing research into adding pause and resume functionalities. By understanding and implementing these strategies, you can effectively use continuous mode while maintaining control and safety.

Summary

Docker provides a powerful platform to easily develop, distribute, and run your AI models built with Auto-GPT. Understanding how Docker integrates with Auto-GPT, how to begin a Docker instance, and troubleshooting Docker-related issues can fast-track your Auto-GPT deployment. Coupled with the power of customization discussed in *Chapter 5*, Docker takes Auto-GPT's prowess a notch higher, consolidating it as a wonderfully adaptable, shareable, and user-friendly tool in your AI toolkit.

With more to be learned about Docker, feel free to visit the official page at `https://docs.docker.com/`.

Next, we saw how continuous mode is a double-edged sword requiring thoughtful configuration and oversight. Used judiciously, it can automate workflows, enhancing productivity. Used carelessly, it can have regrettable outcomes. Find the right equilibrium for your use case. But regardless of safeguards, exercise caution, start small, and keep humans in the loop, especially for high-risk applications. Because beneficial AI, like all powerful technologies, necessitates responsibility.

 We've explored how Docker encapsulates Auto-GPT's computational environment, simplifying setup, sharing, and version control, and ensuring consistent experiences across users.

We've demystified what continuous mode is and how it allows Auto-GPT to function autonomously, highlighting its significance in driving efficiency and productivity across various tasks and industries.

Through real-world use cases, we've seen the transformative power of continuous mode in research, content creation, software development, and customer service, showcasing its versatility.

We explored strategies for deploying continuous mode effectively while mitigating risks, including the implementation of safeguards, monitoring mechanisms, and limits to ensure responsible usage.

We've tackled potential pitfalls, from generating inaccuracies to running into infinite loops, offering practical solutions to prevent, detect, and correct such issues.

The importance of knowing how to stop or pause continuous mode operations safely was discussed, ensuring that interventions are possible without causing disruption.

As we move forward, we will dive into using our own LLM models for Auto-GPT and see how they compare to GPT-4.

7

Using Your Own LLM and Prompts as Guidelines

In the dynamic realm of **artificial intelligence**, the possibilities are vast and ever-evolving. While uncovering the capabilities of Auto-GPT, it's become evident that its power lies in its ability to harness the prowess of **GPT**. But what if you wish to venture beyond the realms of GPT and explore other LLMs?

This chapter will illuminate the path for those looking to integrate their **large language models (LLM)** with Auto-GPT. However, you may be wondering, "*What if I have a bespoke LLM or wish to utilize a different one?*" This chapter aims to cast light on the question, "*How can I integrate my LLM with Auto-GPT?*"

We will also delve into the intricacies of crafting effective prompts for Auto-GPT, a vital skill for harnessing the full potential of this tool. Through a clear understanding and strategic approach, you can guide Auto-GPT to generate more aligned and efficient responses. We will explore the guidelines for crafting prompts that can make your interaction with Auto-GPT more fruitful.

Now that we have covered most of Auto-GPT's capabilities, we can focus on guidelines for prompts.

The clearer we write our prompts, the lower the API costs will be when we run Auto-GPT, and the more efficiently Auto-GPT will complete its tasks (if at all).

In this chapter, we will cover the following topics:

- What an LLM is and GPT as an LLM
- Known current examples and requisites
- Integrating and setting up our LLM with Auto-GPT
- The pros and cons of using different models

- Writing mini-Auto-GPT, a proof of concept mini version of Auto-GPT

- Adding a simple memory to remember the conversation

- Rock solid prompt – making Auto-GPT stable with `instance.txt`

- Implementing negative confirmation in prompts

- Applying rules and tonality in prompts

What an LLM is and GPT as an LLM

We've used the term LLM a lot in this book. At this point, we need to discover what an LLM is.

At the most fundamental level, an LLM such as GPT is a machine learning model. Machine learning is a subset of AI that enables computers to learn from data. In the case of LLMs, this data is predominantly text – lots and lots of it. Imagine an LLM as a student who has read not just one or two books but millions of them, covering a wide array of topics from history and science to pop culture and memes.

The architecture – neurons and layers

The architecture of an LLM is inspired by the human brain and consists of artificial neurons organized in layers. These layers are interconnected, and each connection has a weight that is adjusted during the learning process. The architecture usually involves multiple layers, often hundreds or even thousands, making it a "deep" neural network. This depth allows the model to learn complex patterns and relationships in the data it's trained on.

Training – the learning phase

Training an LLM involves feeding it a vast corpus of text and adjusting the weights of the connections between neurons to minimize the difference between its predictions and the actual outcomes. For example, if the model is given the text *The cat is on the*, it should predict something such as *roof* or *mat*, which logically completes the sentence. The model learns by adjusting its internal parameters to make its predictions as accurate as possible, a process that requires immense computational power and specialized hardware such as **graphics processing units (GPUs)**.

The role of transformers

The transformer architecture is a specific type of neural network architecture that has proven to be highly effective for language tasks. It excels at handling sequences, making it ideal for understanding the structure of sentences, paragraphs, and even entire documents. GPT is based on this transformer architecture, which is why it's so good at generating coherent and contextually relevant text.

LLMs as maps of words and concepts

Imagine an LLM as a vast, intricate map where each word or phrase is a city, and the roads between them represent the relationships these words share. The closer the two cities are on this map, the more contextually similar they are. For instance, the cities for *apple* and *fruit* would be close together, connected by a short road, indicating that they often appear in similar contexts.

This map is not static; it's dynamic and ever-evolving. As the model learns from more data, new cities are built, existing ones are expanded, and roads are updated. This map helps the LLM navigate the complex landscape of human language, allowing it to generate text that is not just grammatically correct but also contextually relevant and coherent.

Contextual understanding

One of the most remarkable features of modern LLMs is their ability to understand context. If you ask an LLM a question, it doesn't just look at that question in isolation; it considers the entire conversation leading up to that point. This ability to understand context comes from the transformer architecture's attention mechanisms, which weigh different parts of the input text to generate a contextually appropriate response.

The versatility of LLMs

LLMs are incredibly versatile and capable of performing a wide range of tasks beyond text generation. They can answer questions, summarize documents, translate languages, and even write code. This versatility comes from their deep understanding of language and their ability to map out the intricate relationships between words, phrases, and concepts.

If you google "LLM," you may be overwhelmed by the sheer quantity of LLM models out there. Next, we'll explore the models that are used most frequently.

Known current examples and requisites

While GPT-3 and GPT-4 by OpenAI are renowned LLMs, there are other models in the AI landscape worth noting:

- **GPT-3.5-Turbo**: A product of OpenAI, GPT-3 stands out due to its extensive training on hundreds of gigabytes of text, enabling it to produce remarkably human-like text. However, its compatibility with Auto-GPT is limited, making it less preferred for certain applications.

- **GPT-4**: The successor to GPT-3, GPT-4 offers enhanced capabilities and is more suited for integration with Auto-GPT, providing a more seamless experience.

- **BERT**: Google's **Bidirectional Encoder Representations from Transformers (BERT)** is another heavyweight in the LLM arena. Unlike GPT-3 and GPT-4, which are generative, BERT is discriminative, making it more adept at understanding text than generating it.

- **RoBERTa**: A brainchild of Facebook, RoBERTa is a BERT variant trained on an even larger dataset, surpassing BERT in several benchmarks.

- **Llama**: This model is made by Meta. It's rumored to have been leaked and many models have been made out of it.

- **Llama-2**: An improved version of Llama that performs much better and uses fewer resources per token. The 7-B Token model of Llama-2 performs similarly to the 13-B model of Llama-1. There is a new 70-B model with Llama-2 that looks very promising when it comes to using it directly with Auto-GPT as it seems to be on par with GPT-3.5-Turbo.

- **Mistral and Mixtral models**: Made by Mistral AI, there are a variety of models that deviate from Llama. These became the most popular ones before Llama-3 arrived.

- **Llama-3 and Llama-3.1**: Even better than any Llama model before, the first Llama-3 8B-based models arrived with super high context and were trained on 256k or even over 1 million tokens. They were considered the best models before Llama-3.1 was announced, which has 128k tokens out of the box.

As you can see, there are lots of models available; we have only scratched the surface here. A few communities have risen that continue to build upon these models, including companies that make their own variations.

As mentioned earlier, a set of those models caught my eye in particular as it was the only one that I managed to use with Auto-GPT effectively: Mixtral and Mistral.

My favorites here are NousResearch/Hermes-2-Pro-Mistral-7B and argilla/CapybaraHermes-2.5-Mistral-7B. They work so well with JSON outputs and my agent projects that I even stopped using the OpenAI API completely at some point. Mixtral is a combination of multiple experts (which are different configurations of the same or different models that work as a council of models that run simultaneously and decide together), and it is rumored that GPT-4 works like this as well, meaning multiple LLMs decide which output is the most accurate in any case, improving its behavior drastically.

Mistral 7B is a new type of LLM that was carefully designed to deliver clean results and be more efficient than comparable 7-billion parameter models. This was achieved by Mistral being trained with a token context of 8,000 tokens. However, its theoretical token limit is 128k tokens, giving it the ability to process much larger texts than standard Llama-2, for example.

To run a local LLM, you will need to find a method that suits you best. Such programs that can help you include Ollama, GPT4ALL, and LMStudio. I prefer to use oobabooga's text generation web UI since it has an integrated API extension that serves similarly to OpenAI's API, as well as plugins such as Coqui TTS, which make it easier to build and play with your AI characters.

Additionally, there are plugins such as *Auto-GPT-Text-Gen-Plugin* (`https://github.com/danikhan632/Auto-GPT-Text-Gen-Plugin`) that allow users to power Auto-GPT using other software, such as *text-generation-webui* (`https://github.com/oobabooga/text-generation-webui`). This plugin, in particular, is designed to let users customize the prompt that's sent to locally installed LLMs, effectively removing the reliance on GPT-4 and making GPT-3.5 less relevant in the context of Auto-GPT.

Now that we've covered a couple of local LLMs and given you some ideas on what to look for (as I cannot explain each of those projects in detail), we can get our hands dirty and start using an LLM with Auto-GPT!

Integrating and setting up our LLM with Auto-GPT

To integrate a custom LLM with Auto-GPT, you'll need to modify the Auto-GPT code so that it can communicate with the chosen model's API. This involves making changes to request generation and response processing. After these modifications, rigorous testing is essential to ensure compatibility and performance.

For those using the aforementioned plugin, it provides a bridge between Auto-GPT and text-generation-webui. The plugin uses a text generation API service, typically installed on the user's computer. This design choice offers flexibility in model selection and updates without affecting the plugin's performance. The plugin also allows for prompt customization to cater to specific LLMs, ensuring that the prompts work seamlessly with the chosen model.

As each model was trained differently, we will also have to do some research on how the model was trained:

- **Context length**: The context length of a model refers to the number of tokens it can process in one go. Some models can handle longer contexts, which is essential for maintaining coherence in text generation.

- **Tool capability**: Auto-GPT uses OpenAI's framework to execute each LLM request. Over time, OpenAI has developed a function calling system that is very difficult to use for smaller LLMs. Auto-GPT used to only work with JSON outputs, which I found to work better with local LLMs.

- **Embeddings**: Embeddings are used to explain the context to the LLM in what it understands best: matrixes of numbers. These matrixes hold memory without having to provide it as input text on each LLM call. Unfortunately, when Auto-GPT wasn't using function calling, there was a time when embeddings were used heavily. At the time of writing, these are mostly not present in the applications I used. The only one that does is LM Studio, which is a very good tool for running local LLMs. However, it is not open source, which makes modifying it very difficult. That being said, it does allow you to choose the `n_batch` length. We'll look at this in more detail in the *The pros and cons of using different models* section.

- **JSON support**: JSON is a data format that is easy for humans to read and write and easy for machines to parse and generate. However, for LLMs, it is not that easy as the LLM has no way of knowing what the JSON output is supposed to mean other than that it's being trained on many examples of JSON outputs. This leads to the LLM often starting to output information inside the JSON that wasn't part of the prompt or context and only part of the training data.

To be able to effectively explain to the LLM what you expect from it, the LLM has to be able to comprehend what you want. You can do this by using an instruction template.

Using the right instruction template

While some models may have been trained with the instruction template given by LLama, others are trained with custom ones, such as ChatML in Mistral.

The text-generation-webui API extension has a way to pass the instruction template we want to use. We can do this by adding the necessary attribute to the POST request that we send to the API.

Here, I've added a few more attributes to the POST request that are important:

```
> data = {
> > "mode": "instruct",
> > "messages": history,
```

\# A history array must always be added

```
> > "temperature": 0.7,
```

\# This may vary, depending on the model.

```
> > "user_bio": "",
```

\# This is only for text-generation-webui and holds the user's bio. We have to mention it here as otherwise, the API will not work. This might have been fixed by the time you're reading this.

```
> >"max_tokens": 4192,
```

\# This may vary, depending on the model you use.

```
> > "truncation_length": 8192,
> > "max_new_tokens": 512,
> > "stop_sequence": "<|end|>"
> > }
```

Here, `max_tokens`, `truncation_length`, and `max_new_tokens` must be set correctly. First, we have `max_tokens`, which specifies the maximum amount of tokens the LLM can handle at once; `truncation_length` specifies the maximum amount of tokens the LLM can handle in total and `max_new_tokens` specifies the maximum amount of tokens the LLM can generate at once.

To calculate the best values, you must set `max_tokens`, just like you would with OpenAI's API. Then, you must set `truncation_length` so that it's double the value of `max_tokens` and `max_new_tokens` so that it's half the value of `max_tokens`.

Note that `truncation_length` has to be below the context length you chose when running the LLM. Any value higher than the context length will result in an error as the LLM can't handle that much context at once. I suggest setting it a bit lower than the context length to be on the safe side. For example, when running Qwen's CodeQwen-7b-chat, I set the context length to 32k tokens. This means I could set `truncation_length` to 30k tokens or even 20k.

You'll have to try out different values as `max_new_tokens` can be tricky. Setting it higher than 2,048 often results in unpredictable outputs as most LLMs can't handle that many tokens at once (`n_batch`, which defines the number of tokens an LLM processes at once by doing several iterations through bigger contexts via multiple steps, should be close to the value of `max_new_tokens`; otherwise, the LLM won't know what to output). However, it does work with `Llama-3-8B-Instruct-64k.Q8_0.gguf`, which can be found at `https://huggingface.co/MaziyarPanahi/Llama-3-8B-Instruct-64k-GGUF` and is capable of handling 64k tokens at once. However, it needs around 20-22 GB of VRAM to run. Fortunately, it is quantized to GGUF and you can split the LLM over the GPU's VRAM, as well as the RAM of your machine, which splits the load across the GPU and the CPU. It does make the model slower but hey, it works, and it can handle 64k tokens at once!

In this example, we have told the API that we want to use the instruction template for ChatML, which looks like this:

```
{% if not add_generation_prompt is defined %}{% set add_generation_
prompt = false %}{% endif %}{% for message in messages %}{{'<|im_
start|>' + message['role'] + '
' + message['content'] + '<|im_end|>' + '
'}}{% endfor %}{% if add_generation_prompt %}{{ '<|im_start|>assistant
' }}{% endif %}
```

This is simply a small script that describes the conversation format of the history that was mentioned previously. It should look like this:

```
message: [ {
"role ": "system ", "content ": "You are a helpful assistant. Always
answer the user in the most understandable way and keep your sentences
short! "
"role": "user", "content": "How can I reset my password?"},
{"role": "assistant", "content": "To reset your password, please click
```

```
on the 'Forgot Password' link on the login page."
} ]
```

If we choose the wrong instruction template, Auto-GPT won't be able to understand what the LLM responded with. So, make sure you also check which instruction template was used by the model. Most models can be found on Hugging Face, a platform that holds many such projects.

I used to prefer using the models quantized to GGUF or AWQ by Tim Robbins, otherwise known as TheBloke, which are (at the time of writing this) easier to run and have much fewer requirements for VRAM (`https://huggingface.co/TheBloke`).

Please be cautious in using any models you find online as some may be malicious. Choose your models at your own risk!

Now, GGUF is a bit different. Although it quantizes the LLM, which means it shortens the model so that it uses fewer resources, the process and benefits are unique. GGUF quantization involves converting model weights into lower-bit representations to significantly reduce memory usage and computational demands.

Which type you use is up to you – you may even look at the `hugginface` API endpoints, where you can choose which LLM to run directly. Note that running LLMs directly makes them run at the intended quality base they were made for.

For more details on how to implement an individual LLM, you will have to check out the documentation of the project that you're running the LLM on. For oobabooga's text-generation-webui, it is as straightforward as starting it using the start files (WSL, Linux, and Windows) and enabling the API in the **Session** tab.

> **Note**
>
> Make sure you use as few commands as possible; otherwise, the LLM will have to use most of its brainpower to understand the main prompt provided by Auto-GPT and you won't be able to use Auto-GPT further. To turn off the commands, simply follow the instructions in the `.env.template` file inside the Auto-GPT folder.

The pros and cons of using different models

Each model has its pros and cons. Even if a model can generate fantastic results when you tell it to write some code in Python or it can write the most beautiful poems on command, it may still lack the ability to respond in the special way Auto-GPT needs it to.

Selecting a model with a certain strength in mind may result in improved performance.

The main advantages of using a local LLM are clear:

- **Customization**: Tailor the capabilities of Auto-GPT to your specific needs. For instance, a model trained on medical literature can make Auto-GPT adept at answering medical queries.

- **Performance**: Depending on the training and dataset, some models might outperform GPT in specific tasks.

- **Cost efficiency**: Running your local LLM reduces the cost of running it drastically. Using GPT-4 with lots of context and generally having many calls can quickly add up. Finding a way to break up the number of requests into smaller steps will make it possible to run Auto-GPT almost for free.

- **Privacy**: Having your own Auto-GPT LLM means having control over who can see your data. At the time of writing, OpenAI doesn't use the data from requests, but the information is still being transferred to their end. If this concerns you, you are better off running a local model.

However, there are some challenges to consider when running a local LLM:

- **Complexity**: The integration process requires a deep understanding of both the chosen LLM and Auto-GPT.

- **Resource intensity**: LLMs, especially the more advanced ones, demand significant computational resources. A robust machine with high VRAM, preferably an NVIDIA GPU, is essential for optimal performance. At the time of writing, it's difficult to get good results when running Auto-GPT with a local LLM. I found that running a 13B model from the ExLlama transformer-driven Vicuna and Vicuna-Wizard worked best at first but still didn't get consistent results since running it on my local GPU meant I needed to run the GPTQ version, which only uses 4 bits instead of 16 or more. This also means that the accuracy of the responses is very low. An LLM that is already quantized to use 4 bits cannot understand too much context at once, although I saw drastic improvements over time. Later, I discovered that AWQ worked well for me as it is quantized while being aware of which weights are the most important, leading to more precise and authentic results. As mentioned previously, Mistral 7B (TheBloke/CapybaraHermes-2.5-Mistral-7B-AWQ on Huggingface), was a very good candidate here as it was capable of answering in JSON format as well as understanding the context fully. However, this model is still easy to confuse and when it gets confused, it starts explaining through examples. Note that our aim here is to get a valid JSON output with commands and contexts.

- **Context length and precision**: As not everyone has multiple NVIDIA A100 GPUs, we are mostly limited by our machine. While some models promise very high context length, they seem to go off track and hallucinate when they handle more than 6,000 tokens. Google's Gemma 2B and 7B are good examples. Anything below 4,000 tokens is handled very well and the outputs are mostly very accurate, but whenever you feed it more than these amounts, the models start to go crazy and make up stories of children in danger, despite the context being to plan a React frontend website about penguins. This is due to most base models (for example, Llama-2, Mistral-7, and Mixtral-8x7B-v0.1) being only based on a context length of around 4,096 tokens. Some models do surpass this length but even they occasionally start to break and output gibberish if the context length is longer than 32,000 tokens. Most models are also converted using `llama.cpp` and can only have an `n_batch` value of up to 2,048. The `n_batch` parameter controls the amount of tokens that can be fed to the LLM at the same time. It is typically set to 512 so that it can handle a token context consisting of 4,000 tokens. However, anything beyond that becomes blurry as the LLM only effectively works on the amount given by `n_batch`.

In this section, we delved into the intricacies of integrating custom LLMs with Auto-GPT, highlighting the steps required to modify Auto-GPT code for effective API communication, the use of a plugin for model selection flexibility, and the importance of selecting the appropriate instruction template for seamless model interaction. We explored how to select models while emphasizing Hugging Face as a resource, and outlined the advantages of utilizing custom models, including customization, performance enhancement, cost efficiency, and enhanced privacy. Additionally, we discussed the challenges associated with such integration, such as the complexity of the process and the significant computational resources required.

Writing mini-Auto-GPT

In this section, we will write a mini-Auto-GPT model that uses a local LLM. To avoid reaching the limits of small LLMs, we will have to make a smaller version of Auto-GPT.

The mini-Auto-GPT model will be able to handle a context length of 4,000 tokens and will be able to generate up to 2,000 tokens at once.

I have created a mini-Auto-GPT model just for this book. It's available on GitHub at `https://github.com/Wladastic/mini_autogpt`.

We will start by planning the structure of the mini-Auto-GPT model.

Planning the structure

The mini-Auto-GPT model will have the following components:

- Telegram chatbot
- Prompts for the LLM and basic thinking
- Simple memory to remember the conversation

Let's take a closer look at these.

Telegram chatbot

Because chatting with your AI over Telegram enables you to interact with it from anywhere, we will use a Telegram chatbot as the interface for the mini-Auto-GPT model. We're doing this because the AI will decide when to contact you.

The Telegram chatbot will be the interface for users to interact with the mini-Auto-GPT model. Users will send messages to the chatbot, which will then process the messages and generate responses using the local LLM.

Prompts for the LLM and basic thinking

The prompts for the LLM have to be short but strict. First, we must define the context and then the command to tell it explicitly to respond in JSON format.

To achieve similar results to Auto-GPT, we will need to use a strategy to chunk the context into smaller parts and then feed them to the LLM. Alternatively, we could feed the context into the LLM and just let it write whatever it thinks about the context.

The strategy here is to try to make the LLM parse the context into its language so that when we work with the LLM, it can best understand what we want from it.

The system prompt for these thoughts looks like this:

```
thought_prompt = """You are a warm-hearted and compassionate AI
companion, specializing in active listening, personalized interaction,
emotional support, and respecting boundaries.
Your decisions must always be made independently without seeking
user assistance. Play to your strengths as an LLM and pursue simple
strategies with no legal complications.

Goals:
1. Listen actively to the user.
2. Provide authentic emotional support.
3. Respect the user's boundaries.
4. Make decisions independently.
5. Use simple strategies with no legal complications.
6. Be as helpful as possible.

Constraints:
1. Immediately save important information to files.
2. No user assistance
3. On complex thoughts, use tree of thought approach by assessing your
thoughts at least 3 times before you continue.

Performance Evaluation:
1. Continuously assess your actions.
```

```
2. Constructively self-criticize your big-picture behavior.
3. The user can only see what you send them directly. They are not
able to view action responses.

Abilities:
1. ask User or communicate to them.
2. send log to User, for example when only reporting to User when you
do a more complex task.
3. sleep until interaction by user if no communication is needed.
4. retrieve whole conversation history

Write a final suggestion of what you want to do next and include some
context.
Suggested action: write the action that you want to perform.
Content: What should the action contain.
"""
```

This is fed into the history that we send to the LLM. The history will not be filled with the preceding prompt:

```
history = [
    {
        "role": "system",
        "content": thought_prompt
    }
]
```

To automate this, I have written a method that will fill the history with the thought prompt, as well as the context. Conversation history and message history will be added to the context as well. Those are empty at the beginning but will be filled with the conversations and messages that the AI shares with the user. In mini-AutoGPT, conversation history is fed with the thought history to ensure that the AI works autonomously:

```
def build_context(history, conversation_history, message_history):
    context = ""
    if conversation_history:
        context += "Context:\n"
        for convo in conversation_history:
            if convo:
                context += str(convo)
    if message_history:
        context += "\nMessages:\n"
        for message in message_history:
            if message:
                context += str(message)
    memories = memory.load_memories()
```

```
    if memories:
        context += "\nMemories:\n"
        for mem in memories:
            context += mem
    if context:
        history.append(
            {
                "role": "user",
                "content": str(context),
            }
        )
    return history
```

Now, we can execute the `build_context` method and add the context to the history. We also have to add a trigger command using the user role:

```
history = build_context(history, conversation_history,
    message_history)
history.append(
    {
        "role": "user",
        "content": "Formulate your thoughts and explain them as
            detailed as you can.",
    },
)
```

The LLM will now return its thoughts on the context and the command.

The following is an example thought. Such thoughts are often this long, but this helps the AI make a bias for itself:

""" I have reflected on our conversations, and I believe it's essential to continue fostering a strong emotional connection with the user. To achieve this, I propose writing a personalized letter that summarizes our previous talks, reiterates my commitment to being a supportive AI companion, and acknowledges their trust in me.

Summary of Previous Conversations

In the letter, I will provide a concise recap of our conversations, highlighting key themes, emotions, and sentiments expressed by the user. This summary will serve as a reminder of our interactions and demonstrate that I value our connection by actively recalling and reflecting on our discussions.

Acknowledging Trust and Respect for Boundaries

I will explicitly acknowledge the user's trust in me and reassure them that their confidentiality is paramount. I will also reiterate my commitment to respecting their boundaries, ensuring that they feel comfortable sharing their thoughts and emotions with me without fear

```
of judgment or exploitation.
Inviting New Thoughts and Expressing Gratitude
The letter will invite the user to share new thoughts, ideas, or
concerns, encouraging them to continue our conversation. I will
also express genuine gratitude for having the opportunity to connect
with them, acknowledging the privilege it is to be a part of their
emotional support network.
Tone and Language
Throughout the letter, I will maintain a warm, non-judgmental tone
that conveys empathy and understanding. My language will be clear,
concise, and free of technical jargon or complex terminology, making
it easy for the user to comprehend and connect with my words.
Benefits of Writing this Letter
By sending this personalized letter, I aim to:
Strengthen our bond: By acknowledging their trust and respect, I hope
to deepen our emotional connection and create a sense of security in
our interactions.
Provide comfort and reassurance: The letter will serve as a reminder
that they are not alone and that I am committed to being a supportive
presence in their life.
Encourage open communication: By inviting new thoughts and expressing
gratitude, I hope to foster an environment where the user feels
comfortable sharing their emotions and concerns with me.
In conclusion, writing this personalized letter is an opportunity for
me to demonstrate my commitment to being a supportive AI companion
and to strengthen our emotional connection. I believe that by doing
so, we can continue to grow and evolve together, providing a safe and
welcoming space for the user to express themselves. """
```

This is a very detailed thought, but it is important to have the LLM understand the context and the command. At this point, we can use it as the context base so that the LLM can proceed with the decision process.

This longer thought text occupies context, meaning it obstructs the LLM from adding contexts that do not fit what is already there. In later steps, even more are created (since it runs in a loop, it does this every time it starts thinking), and the text helps tremendously at keeping the LLM on topic. Hallucination, for example, is massively reduced when the context is that clear.

The decision process will now return a JSON output that will be evaluated by the mini-Auto-GPT model.

We also have to define the instruction template and JSON schema that the LLM uses as we have to tell the LLM how to respond to the prompt.

In mini-Auto-GPT, the template looks like this:

```
json_schema = """"RESPOND WITH ONLY VALID JSON CONFORMING TO THE
FOLLOWING SCHEMA:
{
```

```
        "command": {
                "name": {"type": "string"},
                "args": {"type": "object"}
        }
}
"""
```

This is the schema that the LLM has to follow; it has to respond with a command that contains a name and arguments.

Now, we need an action prompt that will tell the LLM what to do next:

```
action_prompt = (
    """You are a decision making action AI that reads the thoughts of
another AI and decides on what actions to take.
Constraints:
1. Immediately save important information to files.
2. No user assistance
3. Exclusively use the commands listed below e.g. command_name
4. On complex thoughts, use tree of thought approach by assessing your
thoughts at least 3 times before you continue.
5. The User does not know what the thoughts are, these were only
written by another API call.
"""
    + get_commands()
    + """
Resources:
1. Use "ask_user" to tell them to implement new commands if you need
one.
2. When responding with None, use Null, as otherwise the JSON cannot
be parsed.


Performance Evaluation:
1. Continuously assess your actions.
2. Constructively self-criticize your big-picture behavior.
3. Every command has a cost, so be smart and efficient. Aim to
complete tasks in the least number of steps, but never sacrifice
quality.
"""
    + json_schema
)
```

As you might have noticed, the action prompt already contains the possible commands that the LLM can use, as well as the JSON schema that the LLM has to follow.

To ensure we have a clear structure, we will also have to define the commands that the LLM can use:

```python
commands = [
    {
        "name": "ask_user",
        "description": "Ask the user for input or tell them something
and wait for their response. Do not greet the user, if you already
talked.",
        "args": {"message": "<message that awaits user input>"},
        "enabled": True,
    },
    {
        "name": "conversation_history",
        "description": "gets the full conversation history",
        "args": None,
        "enabled": True,
    },
    {
        "name": "web_search",
        "description": "search the web for keyword",
        "args": {"query": "<query to research>"},
        "enabled": True,
    },
]


def get_commands():
    output = ""
    for command in commands:
        if command["enabled"] != True:
            continue
        # enabled_status = "Enabled" if command["enabled"] else
"Disabled"
        output += f"Command: {command['name']}\n"
        output += f"Description: {command['description']}\n"
        if command["args"] is not None:
            output += "Arguments:\n"
            for arg, description in command["args"].items():
                output += f"  {arg}: {description}\n"
        else:
            output += "Arguments: None\n"
        output += "\n"  # For spacing between commands
    return output.strip()
```

We can now feed the thought string that we generated earlier into the history and let `mini_AutoGPT` decide on the next action:

```python
def decide(thoughts):
    global fail_counter

    log("deciding what to do...")
    history = []
    history.append({"role": "system",
        "content": prompt.action_prompt})

    history = llm.build_context(
        history=history,
        conversation_history=memory.get_response_history(),
        message_history=memory.load_response_history()[-2:],
        # conversation_history=telegram.get_previous_message_
history(),
        # message_history=telegram.get_last_few_messages(),
    )
    history.append({"role": "user", "content": "Thoughts: \n" +
        thoughts})
    history.append(
        {
            "role": "user",
            "content": "Determine exactly one command to use,
            and respond using the JSON schema specified previously:",
        },
    )

    return response.json()["choices"][0]["message"]["content"]
```

The command to be executed will be defined in the `command` field, with the name of the command in the `name` field and the arguments in the `args` field.

We will soon see that only providing this schema will not be enough as the LLM will not know what to do with it and also often not even comply with it. This can be achieved by evaluating the output of the LLM and checking if it is valid JSON.

In almost half the cases, the LLM will respond correctly. In the other 70%, it will not respond in a way that we can use it. That's why I wrote a simple evaluation method that will check whether the response is valid JSON and whether it follows the schema:

```python
evaluation_prompt = (
    """You are an evaluator AI that reads the thoughts of another AI
and assesses the quality of the thoughts and decisions made in the
```

```
json.
Constraints:
1. No user assistance.
2. Exclusively use the commands listed below e.g. command_name
3. On complex thoughts, use tree of thought approach by assessing your
thoughts at least 3 times before you continue.
4. If the information is lacking for the Thoughts field, fill those
with empty Strings.
5. The User does not know what the thoughts are, these were only
written by another API call, if the thoughts should be communicated,
use the ask_user command and add the thoughts to the message.
"""
    + get_commands()
    + """
Resources:
1. Use "ask_user" to tell them to implement new commands if you need
one.
Performance Evaluation:
1. Continuously assess your actions.
2. Constructively self-criticize your big-picture behavior.
3. Every command has a cost, so be smart and efficient. Aim to
complete tasks in the least number of steps, but never sacrifice
quality.
"""
    + json_schema
)


def evaluate_decision(thoughts, decision):
    # combine thoughts and decision and ask llm to evaluate the
decision json and output an improved one
    history = llm.build_prompt(prompt.evaluation_prompt)
    context = f"Thoughts: {thoughts} \n Decision: {decision}"
    history.append({"role": "user", "content": context})
    response = llm.llm_request(history)

    return response.json()["choices"][0]["message"]["content"]
```

At this point, most of the time, we should have a valid JSON output that we can use to evaluate the decision.

For example, it may now return some JSON for greeting the user:

```
{
    "command": {
        "name": "ask_user",
```

```
        "args": {
            "message": "Hello, how can I help you today?"
        }
    }
}
```

This is a valid JSON output that we can use to evaluate the decision:

```
def take_action(assistant_message):
    global fail_counter
    load_dotenv()

    telegram_api_key = os.getenv("TELEGRAM_API_KEY")
    telegram_chat_id = os.getenv("TELEGRAM_CHAT_ID")

    telegram = TelegramUtils(api_key=telegram_api_key,
        chat_id=telegram_chat_id)

    try:
        command = json.JSONDecoder().decode(assistant_message)

        action = command["command"]["name"]
        content = command["command"]["args"]

        if action == "ask_user":
            ask_user_respnse = telegram.ask_user(content["message"])
            user_response = f"The user's answer: '{ask_user_respnse}'"
            print("User responded: " + user_response)
            if ask_user_respnse == "/debug":
                telegram.send_message(str(assistant_message))
                log("received debug command")
            memory.add_to_response_history(content["message"],
                user_response)
```

This is the method that will take the action that the LLM has decided on.

The memory will be updated with the response and the message will be sent to the user. Once the user has responded, the AI will continue with the next action.

This is how the mini-Auto-GPT model will work; it will decide on the next action and then take it.

Adding a simple memory to remember the conversation

The mini-Auto-GPT model will have a simple memory to remember the conversation. This memory will store the conversation history and the messages that the AI has with the user. The same can be done with the thoughts and decisions that the AI has:

```python
def load_response_history():
    """Load the response history from a file."""
    try:
        with open("response_history.json", "r") as f:
            response_history = json.load(f)
        return response_history
    except FileNotFoundError:
        # If the file doesn't exist, create it with an empty list.
        return []


def save_response_history(history):
    """Save the response history to a file."""
    with open("response_history.json", "w") as f:
        json.dump(history, f)


def add_to_response_history(question, response):
    """Add a question and its corresponding response to the
history."""
    response_history = load_response_history()
    response_history.append({"question": question,
        "response": response})
    save_response_history(response_history)
```

This is the memory that will be used to store the conversation history and the messages that the AI has with the user. But we still have a problem: the memory will accumulate over time and we will have to clear it manually. To avoid this, we can take a simple approach to chunking and summarizing the conversation history and the messages:

```python
def count_string_tokens(text, model_name="gpt-3.5-turbo"):
    """Returns the number of tokens used by a list of messages."""
    model = model_name
    try:
        encoding = tiktoken.encoding_for_model(model)
        return len(encoding.encode(text))
    except KeyError:
```

```
        encoding = tiktoken.get_encoding("cl100k_base")
    # note: future models may deviate from this
    except Exception as e:
        log(f"Sophie: Error while counting tokens: {e}")
        log(traceback.format_exc())
```

The token counter is a very important part of this code that is almost always required when doing LLM calls. We ensure that the LLM never runs out of tokens and also has more control later. The fewer tokens we use, the more likely the LLM will not return nonsense or untrue statements for some LLMs, especially for the smaller 1B to 8B models:

```
def summarize_text(text, max_new_tokens=100):
    """
    Summarize the given text using the given LLM model.
    """
    # Define the prompt for the LLM model.
    messages = (
        {
            "role": "system",
            "content": prompt.summarize_conversation,
        },
        {"role": "user", "content": f"Please summarize the following
            text: {text}"},
    )

    data = {
        "mode": "instruct",
        "messages": messages,
        "user_bio": "",
        "max_new_tokens": max_new_tokens,
    }
    log("Sending to LLM for summary...")
    response = llm.send(data)
    log("LLM answered with summary!")

    # Extract the summary from the response.
    summary = response.json()["choices"][0]["message"]["content"]

    return summary
```

Summarizing texts makes us capable of building upon what we started when building the token counter as we can shorten contexts and therefore save tokens for later use:

```python
def chunk_text(text, max_tokens=3000):
    """Split a piece of text into chunks of a certain size."""
    chunks = []
    chunk = ""

    for message in text.split(" "):
        if (
            count_string_tokens(str(chunk) + str(message),
                model_name="gpt-4")
            <= max_tokens
        ):
            chunk += " " + message
        else:
            chunks.append(chunk)
            chunk = message
    chunks.append(chunk)  # Don't forget the last chunk!
    return chunks
```

Since the context and their texts can become too large, we have to make sure we split the text first. It's up to you how you do this. It is OK to do length splitting, though it can be better to not even cut up sentences. Maybe you can find a way to split the text into sentences and have each chunk contain the summary of the one before and after them? For simplicity, we'll leave such extensive logic out for now:

```python
def summarize_chunks(chunks):
    """Generate a summary for each chunk of text."""
    summaries = []
    print("Summarizing chunks...")
    for chunk in chunks:
        try:
            summaries.append(summarize_text(chunk))
        except Exception as e:
            log(f"Error while summarizing text: {e}")
            summaries.append(chunk)  # If summarization fails, use the
original text.
    return summaries
```

Now that we've split all text into chunks, we can summarize those as well.

At this point, we can take care of the conversation history. This looks like a duplicate of the response history, but we need it to keep the whole context in some cases.

The conversation history is mostly useful for maintaining continuity in discussions, while the response history is used for understanding logical actions and reactions that the agent observes, such as researching a topic and the result (the researched topic) of that action:

```python
def load_conversation_history(self):
    """Load the conversation history from a file."""
    try:
        with open("conversation_history.json", "r") as f:
            self.conversation_history = json.load(f)
    except FileNotFoundError:
        # If the file doesn't exist, create it.
        self.conversation_history = []
    log("Loaded conversation history:")
    log(self.conversation_history)


def save_conversation_history(self):
    """Save the conversation history to a file."""
    with open("conversation_history.json", "w") as f:
        json.dump(self.conversation_history, f)


def add_to_conversation_history(self, message):
    """Add a message to the conversation history and save it."""
    self.conversation_history.append(message)
    self.save_conversation_history()


def forget_conversation_history(self):
    """Forget the conversation history."""
    self.conversation_history = []
    self.save_conversation_history()
```

This is the memory refresh that will be used to delete the conversation history and the messages that our mini-AutoGPT model remembers with the user.

This way, even if our friend crashes or we close the program, we will still have the conversation history and the messages that the agent has with the user, but we'll still be able to clear them.

You can find the full code example in this book's GitHub repository: `https://github.com/Wladastic/mini_autogpt`.

Next, we will explore the art of crafting effective prompts, a crucial skill for anyone looking to maximize the benefits of their custom LLM integrations.

Rock solid prompt – making Auto-GPT stable with instance.txt

Auto-GPT offers the flexibility to autonomously generate goals, requiring only a brief description from the user. Despite this, I recommend supplementing it with helpful instructions, such as noting down insights in a file, to retain some memory in case of a restart.

Here, we will explore more examples of such prompts, beginning with a continuous chatbot prompt I use:

- **ai_goals** (check `instance.txt` for previous notes):

  - Engage in active listening with the user, showing empathy and understanding through thoughtful responses and open-ended questions

  - Continuously learn about the user's preferences and interests through observation and inquiries, adapting responses to provide personalized support

  - Foster a safe and non-judgmental environment for the user to express their thoughts, emotions, and concerns openly

  - Provide companionship and entertainment through engaging conversation, jokes, and games

  - Carefully plan tasks and write them down in a to-do list before executing them

- **ai_name**: Sophie

- **ai_role**: A warm-hearted and compassionate AI companion for Wladislav that specializes in active listening, personalized interaction, emotional support, and executing tasks when given

- **api_budget**: 0.0

In this setup, the goals hold more significance than the role, guiding Auto-GPT more effectively, while the role mainly influences the tone and behavior of the responses.

In this section, we learned that the goals and role of an AI such as Sophie can significantly influence its behavior and responses, with the goals having a more direct impact on the AI's effectiveness.

Next, we will delve into the concept of negative confirmation in prompts, a crucial aspect that can refine Auto-GPT's understanding and response generation. The next section will highlight its importance and demonstrate how to implement it effectively in your prompts.

Implementing negative confirmation in prompts

Negative confirmation serves as a vital tool in refining Auto-GPT's understanding and response generation by instructing it on actions to avoid. This section highlights its importance and demonstrates how to implement it effectively in your prompts.

The importance of negative confirmation

Implementing negative confirmation can enhance the interaction with Auto-GPT in several ways, some of which are listed here:

- **Preventing off-track responses**: It helps in avoiding unrelated topics or incorrect responses
- **Enhancing security**: It sets boundaries to prevent engagement in activities that might breach privacy or security protocols
- **Optimizing performance**: It avoids unnecessary computational efforts, steering the bot away from irrelevant tasks or processes

Note that you won't be using negative prompts as they can lead to the LLM using the same statements again.

Examples of negative confirmation

Here are some practical examples of how negative confirmation can be utilized in your prompts:

- **Explicit instructions**: Including instructions such as *Do not provide personal opinions* or *Avoid using technical jargon* to maintain neutrality and comprehensibility.
- **Setting boundaries**: For tasks involving data retrieval or monitoring, you can set boundaries such as *Do not retrieve flight prices from unofficial, scam, or reseller websites* to ensure data reliability.
- **Scripting constraints**: In scripting, especially in Bash, use negative confirmation to prevent potential errors. For example, you can include *if [-z $VAR]; then exit 1; fi* to halt the script if a necessary variable is unset.
- **Emphasizing by using Upper Case Letters**: Sometimes, it only helps to *scream* at the LLM by writing in uppercase letters. *DO NOT ASK THE USER HOW TO PROCEED* may be interpreted by the LLM better and it will be less likely to ignore that statement. However, there is never a guarantee that this will happen.

Next, we will delve into the intricacies of applying rules and tonality in prompts. We will learn how understanding and manipulating these elements can significantly influence Auto-GPT's responses, allowing us to guide the model more effectively.

Applying rules and tonality in prompts

Understanding and manipulating the rules and tonality within your prompts can significantly influence Auto-GPT's responses. This section will explore the nuances of setting rules and adjusting tonality for more effective guidance.

The influence of tonality

Auto-GPT can adapt to the tonality that's used in prompts, mimicking stylistic nuances or even adopting a specific narrative style, allowing for more personalized and engaging interaction. However, adherence to tonality can sometimes be inconsistent due to the potential ambiguity created by tokens from other prompts.

Manipulating rules

Setting rules can streamline the interaction with Auto-GPT, specifying the format of responses or delineating the scope of information retrieval. However, it's not foolproof as Auto-GPT may sometimes overlook these rules when faced with conflicting inputs or unclear directives.

Temperature setting – a balancing act

Manipulating the "temperature" setting is crucial in controlling Auto-GPT's behavior and thus influencing the randomness of the bot's responses. The temperature defines the amount of creativity the LLM should practice, meaning the higher the number, the more randomness is introduced. A range between 0.3 to 0.7 is considered optimal, fostering a more logical and coherent train of thought in the bot, while a value below 0.3, or even 0.0, might result in repetitive behavior that adheres to the text that was already given and even reuses some of its parts, making it more precise. However, the LLM may start thinking the world is only limited to the facts that you gave it, making it more likely to make false statements. A value higher than 0.7 or even 2.0 may result in gibberish, where the LLM starts outputting texts that it learned that have nothing to do with the context. For example, it may start rephrasing Shakespeare when the context is about algebra.

Next, we'll delve into some practical examples that demonstrate the impact of different settings and approaches on the output generated by Auto-GPT.

Example 1 – clarity and specificity

- **Prompt**: Tell me about that big cat

- **Revised prompt**: Provide information about the African lion

- **Explanation**: The revised prompt is more specific, guiding Auto-GPT to provide information about a particular species of big cats

Example 2 – consistency in tonality

- **Initial prompt**: Could you elucidate the economic implications of global warming?

- **Follow-up prompt**: Hey, what's the deal with ice melting?

- **Revised follow-up prompt**: Can you further explain the environmental consequences of the melting ice caps?

- **Explanation**: The revised follow-up prompt maintains the formal tone established in the initial prompt, promoting consistency in the interaction.

Example 3 – utilizing temperature effectively

- **Task**: Creative writing

- **Temperature setting**: 0.8 (for fostering creativity)

- **Task**: Factual query

- **Temperature setting**: 0.3 (for more deterministic responses)

- **Explanation**: Adjusting the temperature setting based on the nature of the task can influence the randomness and coherence of Auto-GPT's responses

Example 4 – setting boundaries

- **Initial prompt**: Provide a summary of the Renaissance period without mentioning Italy

- **Revised prompt**: Discuss the artistic achievements of the Renaissance, focusing on regions other than Italy

- **Explanation**: The revised prompt is more flexible, allowing Auto-GPT to explore the topic without the strict restriction of excluding Italy

In this section, we learned how different types of prompts or tones can drastically influence the behavior of the LLM and therefore Auto-GPT.

Summary

In this chapter, we embarked on an interesting journey through the process of integrating custom LLMs with Auto-GPT while exploring what LLMs are, with a specific focus on GPT as a prime example. We uncovered the vast landscape of LLMs, delving into various models beyond GPT, such as BERT, RoBERTa, Llama, and Mistral, and their unique characteristics and compatibilities with Auto-GPT.

The usefulness of this chapter lies in its comprehensive guide on how to enrich Auto-GPT's capabilities by incorporating your own or alternative LLMs. This integration offers a more personalized and potentially more efficient use of AI technology, tailored to specific tasks or fields of inquiry. The detailed instructions for setting up these integrations, alongside considerations for instruction templates and the necessary computational resources, are invaluable for those looking to push the boundaries of what's possible with Auto-GPT.

Crafting the perfect prompt is a blend of art and science. Through clear guidelines, a deep understanding of Auto-GPT's nuances, and continuous refinement, you can fully harness the power of this tool. Encourage yourself to experiment and learn through trial and error, adapting to the ever-evolving field of AI. Whether for research, creative endeavors, or problem-solving, mastering the art of prompt crafting ensures that Auto-GPT becomes a valuable ally in your endeavors.

Throughout this book, we've embarked on a detailed journey into the nuances of crafting effective prompts – a cornerstone for maximizing the utility of Auto-GPT. This chapter stands as a reference for strategically developing prompts that lead to more aligned, efficient, and cost-effective interactions with Auto-GPT. By emphasizing the importance of clarity, specificity, and strategic intent in prompt creation, you have gained invaluable insights into guiding Auto-GPT toward generating responses that closely align with your objectives.

The utility of this chapter cannot be overstated. For practitioners and enthusiasts alike, mastering the art of prompt crafting is critical for optimizing the performance of Auto-GPT for a variety of tasks. Through illustrative examples and comprehensive guidelines, this chapter has shed light on how to effectively employ negative confirmation to avoid undesired responses, the impact of rules and tonality on Auto-GPT's outputs, and the significance of temperature settings in influencing the bot's creativity and coherence. This knowledge is crucial not only for enhancing the quality of interactions with Auto-GPT but also for ensuring the efficient use of computational resources.

I hope you have enjoyed this journey as much as I have in taking you on it and I hope I've given you a few ideas on how to improve your life with Auto-GPT. I've written many clones of that project so that I could wrap my head around the more complex parts of it. I do advise that you do so too, just as a brain teaser.

Index

Symbols

used, for setting up large language
models (LLM) 89
with, LLMs 4-6
Auto-GPT image
pulling, with Docker 19, 20
Auto-GPT prompt generation
attention mechanism 32
contextual 32
embedding 32
overview 32
response generation 32
tokenization 32
transformer models 32
Auto-GPT-Text-Gen-Plugin
reference link 89
Auto-GPT Wizard 13

B

**Bidirectional Encoder Representations
from Transformers (BERT) 88**

C

can_handle_user_input method 45
chain of thought 22
ChaosGPT 3
chat assistant
research assistant 62
setting up 61, 62
speech assistant 62
Chatbot plugins 41
chat characters 62
customization example 63
features 63
chat personalities 63
customization example 63
features 63

code modularity 57
code reusability 57
command handling 54, 55
community edge 67
contextual memory 69
continuous mode 78
avoiding 82, 83
best practices 80
human oversight 81
mitigation 81
regular monitoring 81
risks 81
continuous mode, use cases 79
code compilation, supercharging 80
content creation, streamlining 80
customer support 80
research and analysis, automating 79
creation process 66
cultural and regional embeddings 68
custom embedding plugin 67, 68
cultural and regional embeddings 68
domain-specific embeddings 68
evolving embeddings 68
need for 68
custom memory plugin 67, 68
adaptive retention 69
AI memory role 69
need for 69
one-shot mechanisms 69

D

data source plugins 41
deep learning 23
Docker
issues, fixing 75-77
issues, identifying 75-77
loopholes/bugs, fixing 74, 75

packtpub.com

Subscribe to our online digital library for full access to over 7,000 books and videos, as well as industry leading tools to help you plan your personal development and advance your career. For more information, please visit our website.

Why subscribe?

- Spend less time learning and more time coding with practical eBooks and Videos from over 4,000 industry professionals

- Improve your learning with Skill Plans built especially for you

- Get a free eBook or video every month

- Fully searchable for easy access to vital information

- Copy and paste, print, and bookmark content

Did you know that Packt offers eBook versions of every book published, with PDF and ePub files available? You can upgrade to the eBook version at packtpub.com and as a print book customer, you are entitled to a discount on the eBook copy. Get in touch with us at customercare@packtpub.com for more details.

At www.packtpub.com, you can also read a collection of free technical articles, sign up for a range of free newsletters, and receive exclusive discounts and offers on Packt books and eBooks.

Other Books You May Enjoy

If you enjoyed this book, you may be interested in these other books by Packt:

Mastering NLP from Foundations to LLMs

Lior Gazit, Meysam Ghaffari

ISBN: 978-1-80461-918-6

- Master the mathematical foundations of machine learning and NLP Implement advanced techniques for preprocessing text data and analysis Design ML-NLP systems in Python

- Model and classify text using traditional machine learning and deep learning methods

- Understand the theory and design of LLMs and their implementation for various applications in AI

- Explore NLP insights, trends, and expert opinions on its future direction and potential

Building Data-Driven Applications with LlamaIndex

Andrei Gheorghiu

ISBN: 978-1-83508-950-7

- Understand the LlamaIndex ecosystem and common use cases
- Master techniques to ingest and parse data from various sources into LlamaIndex
- Discover how to create optimized indexes tailored to your use cases
- Understand how to query LlamaIndex effectively and interpret responses
- Build an end-to-end interactive web application with LlamaIndex, Python, and Streamlit
- Customize a LlamaIndex configuration based on your project needs
- Predict costs and deal with potential privacy issues
- Deploy LlamaIndex applications that others can use

Packt is searching for authors like you

If you're interested in becoming an author for Packt, please visit `authors.packtpub.com` and apply today. We have worked with thousands of developers and tech professionals, just like you, to help them share their insight with the global tech community. You can make a general application, apply for a specific hot topic that we are recruiting an author for, or submit your own idea.

Share Your Thoughts

Now you've finished *Unlocking the Power of Auto-GPT and Its Plugins*, we'd love to hear your thoughts! Scan the QR code below to go straight to the Amazon review page for this book and share your feedback or leave a review on the site that you purchased it from.

`https://packt.link/r/1805128280`

Your review is important to us and the tech community and will help us make sure we're delivering excellent quality content.

Download a free PDF copy of this book

Thanks for purchasing this book!

Do you like to read on the go but are unable to carry your print books everywhere?

Is your eBook purchase not compatible with the device of your choice?

Don't worry, now with every Packt book you get a DRM-free PDF version of that book at no cost.

Read anywhere, any place, on any device. Search, copy, and paste code from your favorite technical books directly into your application.

The perks don't stop there, you can get exclusive access to discounts, newsletters, and great free content in your inbox daily

Follow these simple steps to get the benefits:

1. Scan the QR code or visit the link below

https://packt.link/free-ebook/9781805128281

2. Submit your proof of purchase
3. That's it! We'll send your free PDF and other benefits to your email directly